Real Physics and Cosmology

The Major scientific breakthrough since Galileo!

All phenomena have a mechanical source!

Bye-bye all no-aether theories!

Josef Tsau

ISBN 0-7414-3851-8

Published by:

INFINITY
PUBLISHING.COM

1094 New DeHaven Street, Suite 100
West Conshohocken, PA 19428-2713
Info@buybooksontheweb.com
www.buybooksontheweb.com
Toll-free (877) BUY BOOK
Local Phone (610) 941-9999
Fax (610) 941-9959

Printed in the United States of America

Printed on Recycled Paper

Published March 2007

To real science and my parents

This book is dedicated to the only real nature science for the welfare and the future of mankind. It is also for the memory of my parents, who brought me up to be a dedicated scientist.

Preface

The author's breakthrough discovery of both aether and new aether science about nine years ago has expanded physics of mechanics to all universal phenomena, thus, becoming the science of everything. It has been a hard learning process for him to enter and to understand such a wonderful new science world. He therefore has to continue to write new book to replace the old. This book is the fourth replacing "Theory of the Universe" published in 1999, "Mysteries of Nature" in 2002, "Discovery of Aether and its Science" in 2005. He is finally satisfied with that all important known universal phenomena are now included and hope that this book will be his last. Although important additions and some revisions have been made from book to book, the fundamental discovery that the atmosphere of neutrinos is the common scientific origin of all universal phenomena has continuously been confirmed. Like putting all the pieces of a puzzle game together, the discovered new aether science offers a broad scientific understanding of all universal phenomena and the universe, thus, surely establishing itself joining physics of mechanics as the only real nature science.

The book addresses the following important scientific questions, which have been asked for centuries. Is there aether in space to transmit light and what is it? Is there simple common-sense-making scientific interpretation on gravity, universal forces, light, etc.? What are charges and why elemental particles are permanently charged? Why protons and neutrons form nuclei, atoms and molecules with electrons? Why electrons in atoms and molecules take up quantized orbital and do not fall into nuclei? What are wrong with theory of relativity, big bang theory, and all other no-aether theories? What is a universe and where it came from? Why planets in solar system have steady-state orbits and rotational motions? How stars form and how can they start and maintain nuclear reactions at essentially constant rates at their centers for probably billions of years? Why so many galaxies have spiral shapes and are rotating? What black hole, neutron stars, white dwarfs really are?

Please find the answers to the above scientific questions and a lot more in this little book! Compare them to what you have learnt and it should be easy for you to determine what the only real science is. This book has three very important missions: to introduce new breakthrough scientific discoveries, to convince the world that the physics of mechanics Galileo has led to develop is the only real science and that all no-aether theories including modern physics and modern cosmology are wrong.

Scientists have been dreaming for a simple common-sense-making coherent scientific understanding of the nature for centuries. It is now here and is supported by huge amount of experimental findings and observations accumulated over four centuries. It has a scope broader than all the no-aether theories combined. Yet, it fits so well together like putting all the pieces of a difficult jigsaw-puzzle game together leaving the possibility of being wrong essentially zero. It is even a lot better than Einstein's dream of a theory of everything since it is both not a postulations-based theory and consistent with physics of mechanics, the only part of the physics having been repeatedly proven correct without unproved postulation.

Nature science was associated with religious beliefs and was a religious pseudo-science from the very beginning. The geocentric era had lasted for well over a thousand years. Copernicus and his followers in 16th Century had started a scientific revolution to separate nature science from religion. His famous follower Galileo in 17th Century had also led to search for real science and discover physics of mechanics and scientific method of using experimental and observational findings, accumulated scientific knowledge, and logic thinking to make scientific discovery. Physics of mechanics has since been expanded and repeatedly proven correct over the past four centuries and long become the core of physics and "the only proven real physics" having no unproved postulation. Physics of mechanics has shown that many phenomena (forces and/or energies) were produced by matters having masses and relative motions and predicted that all phenomena were produced by matters. Nature science therefore is to study matters and the phenomena they produce.

For several centuries, the scientific revolution of Copernicus and Galileo has encountered a major scientific difficulty: the failure to find the matters (aether) predicted by physics of mechanics to produce universal phenomena. The failure had long led to the fatal mistake of violating its teachings to develop and to accept no-aether theories. Finally, Einstein's (un)scientific revolution a century ago has led the entire scientific community worldwide to openly support no-aether theories and deny physics of mechanics and its scientific method. Again, the world has fallen into a new religious pseudo-science era for a century already, which is amazingly similar to that of geocertrism since both relying on unscientific postulations (religious beliefs) and mathematical models.

Upon the discovery of aether this book has expanded physics of mechanics to the science of everything and disproved all no-aether theories, which should not be acceptable to begin with. Upon reading this book you have been given the mission to continue the scientific revolution of Copernicus and Galileo. Although today's enemies of science, the scientific community and the scientific agencies of the governments, are much stronger than church, they are however obligated to support real science not religious pseudo-science. You have the right to challenge them and demand them to face and to correct the biggest scientific mistake accumulated over several centuries. You also have the right to sponsor public debates to determine what the only real nature science should be. You even have the right to seek legal action to stop the corruptions of scientific communities and government agencies from using our tax dollars to support religious pseudo-science in the name of real science.

It is time for the public including all scientists to learn new aether science, to help ending the new era of religious pseudo-science, and to bring our world back to real science. This is not a hopeless task since the new aether science makes common sense and is easy to understand and so is the scientific issue: there can be only one real nature science.

Table of Contents_Toc156264935

Chapter 1 – The Only Real Nature Science

A. Nature Science and Scientific Method 1
B. Physics of Mechanics (PM)..3
Core of Physics 3
The Fundamental (Having-Aether) Science 4
Excluding Newton's Universal Theory of Gravitation 5
Standing Tall 6

Chapter 2 – Universal Phenomena

A. Discovery of Aether 7
B. Mechanical Gravity 8
Scientific Force Fields 8
Scientific (Mechanical) Gravity 9
Unique Properties of Mechanical Gravity 12
Mechanical Gravity V. Spacetime Gravity 15
Drag Effect and Planetary Motions 18
C. Other Universal Forces and Energies 22
Strong Nuclear Force 23
Weak Force 24
Charged Particles, Electric and Magnetic Forces 25
Heat and Temperature 30
A Revisit of Lesage Gravity 32
D. Mechanical Light and its Universe....................................34
Misconceptions 34
Scientific Origin of Light 39
Doppler Effects 40
Universal Red-Shift and Big Bang Theory 42
Scientific Interpretation of Universal Red-Shift..............44
Energy Source and Backgrounds of Light 46
Neutrino Winds and CMB...47

Chapter 3 – Cosmology

A. Fundamental Universal Views 51
The Domain of the Nature (DON) 51

Open Universe 53
Dense-Matter Objects 53
Formation of a Galaxy and a Universe.............................55

B. Stars 57
Astronomical Findings and Modern Cosmology 57
The Stars of the New Aether Science................................67

Chapter 4 – Particle Physics

A. Modern Physics 75
B. New Aether Science 77
Absolutely Uncharged Particles..77
The Universe of Protons and Electrons............................78
The Universe of Neutrinos...79
Nuclear Energy...80

Chapter 5 –Problems in Physical Science

A. Strong Signs of Serious Scientific Problems.......................81
B. The Geocentric Era - The Wrong Beginning.......................86
C. The Rise of No-Aether Theories..87
Newton's Theory of Universal Gravitation.......................88
Faraday's Force Field Theory...89
Maxwell's Light Theory 89
Modern Physics 90
Incompatibility with PM..100
D. The New Era of Religious Pseudo-Science.......................101
E. The Only Scientific Solution 105

Chapter 6 – The Scientific Revolution

A. The Purposes of the Scientific Revolution 106
To separate Science from Religion 106
To Discover Real Science...107
B. The Comeback of Religious Pseudo-Science 108
C. Fighting for Real Science 109
The Corrupted Physical Science Journals 110
National Science Foundation...110
Legal Challenge...119

Chapter 1 – The Only Real Nature Science

A. Nature Science and Scientific Method

Nature science is the laws of the nature, which are experimentally discoverable only. It can neither be changed nor be man-made. Scientific method uses experimental findings and observations to discover nature laws. All animals have the ability to adopt some laws of the nature to survive. Human, however, has been given the most powerful logic thinking ability to understand nature science. Scientific method also includes using logic thinking to interpret findings and observations, to figure out their technological applications, and to discover new natural laws. Since there can be only one correct scientific answer for everything, nature science must be consistent without contradiction. The scientific knowledge therefore must be accumulative to maintain consistency. Scientific method therefore further involves obeying the proven-correct scientific laws and knowledge and using them as the guide to further advance nature science ensuring consistency.

Due to the consistency and the accumulative nature, it is most important that the foundation of science is correct. Otherwise, the entire science would be wrong. For example, the well known geocentric cosmology before 17th Century was based on the religious belief or unscientific postulation that Earth was the center of the universe. It, including all of its mathematical models, is a completely wrong science. It was very difficult to correct both scientifically and socially and had lasted over a thousand years since it had become religion-like. To advance science consistently, proven-correct scientific laws and knowledge must be repeatedly challenged but must also be obeyed. Scientific laws and knowledge must be repeatedly tested for a long period of time to assure its correctness. The postulations of scientific theories must obey proven-correct laws and knowledge of science to assure their scientific correctness and overall scientific consistency. Therefore, a theory based on postulations inconsistent with proven correct science should not be acceptable even if it is supported by some experimental findings. It is like to obey rules to play and to stay in any game. However, unlike nature science the rules of games are man-made, which can be changed, e.g. with the agreement of the players. The rules of the nature science however can only be discovered, proved or disproved using scientific method. It is the obligation of scientific societies, authorities and individual scientists to obey and to apply the correct knowledge and laws of science to advance science.

The most important question is, therefore, what is the correct science? The best answer is that it must be supported by all experimental findings and observations. It has been over four centuries since the beginning of scientific research and an amazingly huge amount of experimental findings and observations have been generated and accumulated. Only the correct or real science can be supported by all of them. There has been little doubt on what is the real science in almost all scientific fields except physics and cosmology. It is time to determine what the only real physics and cosmology are.

Many experimental findings may be interpreted differently and therefore all valid scientific interpretations must be considered. For example, the "well-established scientific knowledge" for nuclear reaction is that a fraction of mass is lost turning into huge amounts of (pure) energy (obeying Einstein Equation). This interpretation however contradicts the well-established mass-balance law of the physical and the chemical reactions. The same experimental findings however can also be interpreted as: the mass lost in all nuclear reactions is due to the release of mass-containing particles such as neutrinos, which are too small to be detectable by available technologies. This interpretation does not violate the well established mass-balance law and the huge amounts of experimental findings supporting the law. Having not considered or unscientifically ruling out this possible scientific interpretation, the above so-called "well established scientific knowledge" only has 50% chance to be correct. Worse yet, for it to be scientifically valid, its term of "pure energy" must be both scientifically definable and obey the mass-balance law. According to well-established scientific knowledge, all known energies are produced by matters having masses and (relative) motions. "Pure energy" without any matter to produce it is scientifically indefinable or unscientific. The above "well-established knowledge" of nuclear reactions therefore must be scientifically wrong and must be replaced by the above proposed new scientific interpretation that is supported by all experimental findings!

B. Physics of Mechanics (PM)

Core of Physics

In addition to his important contribution to astronomy, Galileo is the founder of physics of mechanics (PM). Scientific physics therefore is considered to begin from Galileo in 17th century. PM studies motions of objects and motions related fundamental physical properties of matters, such as momentum, energies, and forces. Newton (1642-1727), the successor of Galileo, became famous for his three laws of motion and his universal theory of gravitation. By 19th Centuries PM has been expanded to cover physics of heat, thermodynamics and kinetics. Heat is the study of energy, energy transfer and heat phenomena. Heat phenomenon has been proven to be the mechanical energy of atoms and molecules and temperature is a measure of the kinetic energy of atomic and molecular motions. PM has long been recognized to be the core of physics. It is also often referred to as the determinant physics meaning that it is experimentally verifiable and quantifiable. It is also the part of physics that makes common sense emphasizing logical understanding. However, its progress has been seriously hindered from the long-time failure in finding the matters (aether) it has predicted to produce such universal phenomena as light and gravity.

PM has been and remains to be the only well-established part of the physics that has been repeated proven correct over the past four centuries without unproved postulations. Even today PM is still recognized to be the core physics. We should be very sure by now that PM is the correct science. Physics is the study of the basic properties of matters (objects) and therefore is the fundamental science of all branches of sciences. It is therefore particularly important that physics is correct science.

The Fundamental (Having-Aether) Science

PM is the having-aether science and therefore it contradicts all no-aether theories! Yet, PM has long been the fundamental physics taught in schools and colleges along with no-aether theories such as force field theory, electromagnetism and modern physics. The PM we learn in schools and colleges however emphasizes mathematical applications not its fundamental science. This serious deemphasize or omission of teaching the fundamental science of PM reflects the fact that physicists have long lost confidence in the science of PM and had turned around to support no-aether theories. Also, they have been doing so to avoid facing serious scientific contradiction problems in physics.

According to PM, there are both matters and phenomena in space and time. All matters have masses, occupy certain amounts of space, and have relative motions. As a result of having these basic physical properties, only matters can produce forces and energies. Matters therefore have relative energy levels and they sometimes undergo collision interactions with each other when some of them happen to enter the same spaces at the same time. Objects produce forces to act on each other and to exchange energies in their collision interactions resulting in changing their relative motions and energy levels. Like having mass and relative motions, having momentum, energies, and forces are part of the fundamental physical properties of all matters. This scientific fact is also understandable mathematically since all the above-mentioned basic properties of objects have been defined mathematically in terms of the masses and the motions (velocity with or without acceleration) of matters. PM therefore teaches that (1) only matters can produce momentum, forces and energies (2) matters produce "pushing" forces acting on contact only. Since all phenomena have energies and/or forces, PM further teaches that (3) Objects should produce all phenomena. The third conclusion has been confirmed by many phenomena such as fire, atmospheric pressure, water wave, sound, etc. Physicists therefore have long predicted the presence of aether (matters) taking up all space to produce and transmit light and universal forces but they have failed to find it for centuries.

Since PM has been repeatedly proven correct over the past several centuries, its teachings must be obeyed. Scientists, therefore, have no choice but continue to search for aether. The scope of scientific research has therefore been narrowed down by the teachings of PM to be the study of matters in space and time, since all phenomena are produced by them. The scientific origins of phenomena, such as fire, and sound, therefore, must be an object, which can only be discoverable experimentally. A phenomenon is not a mathematical function and cannot be discovered by mathematical manipulations.

Air particles produce such worldwide phenomena as wind, atmospheric pressure, fire, and sound. PM teaches that universal phenomena, such as universal forces and light, should also be produced by matters(s) (aether).

Excluding Newton's Universal Theory of Gravitation

Both Newton's three laws of motion and theory of universal gravitation have been considered to be important parts of PM. However, his theory of universal gravitation theory is actually a no-aether theory, not PM. It is not even consistent with PM. It is based on such no-aether postulations as that all matters have acting-from-distance attraction force. It is not even scientifically compatible with PM, the having-aether science. Without Newton's universal theory of gravitation, PM has no unproved postulations and is therefore a proven correct science not a theory.

Standing Tall

There is no way to support both PM and no-aether theories due to there fundamental contradiction. The having-aether science of PM has long been officially banned for applying to universal phenomena. Both the physical science communities and the scientific authorities have long supported no-aether theories and turned their back on PM, the correct (real) science.

Despite what physicists and their authorities believe and what schools teach, experimental findings are the only way to reveal what real science is. Although some experimental results have more than one interpretation, very large amounts of experimental findings always point to the only correct science. In the presence of very large amount of findings, the only real science will be found standing out supported by all experimental findings. Over the past four centuries, tremendous amounts of experimental findings have been generated. By carefully reviewing all of them; the only correct or real science will stand tall and cannot be denied. Certainly, it is the case for PM, which has been repeatedly proven correct scientifically, mathematically, and in all technological applications over the past centuries.

Even after a century into the era of pseudo-physical science, PM is still the undeniable fundamental physics taught in schools and colleges. Although physicists have long been alleging that PM has a valid range, all of their statements are based on no-aether theories. Now the most important prediction of PM that there must be matters (aether) in space to produce all phenomena including universal forces and light has also been proven correct by discovering aether and its science. The discovery has finally expanded PM to the science of everything and is now ready to replace all no-aether theories including modern physics.

Upon finding aether now all universal forces are explainable as act-on-contact pushing forces, thus, proving again that only PM's teachings are supported by all experimental findings.

Chapter 2 – Universal Phenomena

A. Discovery of Aether

Scientists have long discovered that many phenomena, such as sound, fire, atmospheric pressure, wind, even the presence of animal and plants, etc. can occur anywhere on Earth, are mainly or partially due to the presence of an atmosphere of air. For phenomena such as universal forces and light to occur anywhere in the universe, there should be an atmosphere of something (aether) taking up all the space in the universe to produce and to transmit them.

Earth does not continue to produce air and the atmosphere of air is kept on the surface of Earth by gravitational force. Without gravity air will quickly escape from Earth. The nature appears to have unlimited space and we may define our universe to be a region where many stars exist. If there are gas-like particles in the universe (aether), they too will continuously escape out of the universe, where no star exists. For the universe to have an atmosphere of aether having essentially constant concentration and energy to produce and transmit all universal phenomena, the universe must have a way to continuously produce them at a constant rate to compensate their loss from escaping out of the universe. Besides, there must be many locations about evenly distributed in the universe to produce aether in order to have an essentially even concentration and energy distribution of aether in the universe. Adding all the above difficult scientific conditions together, there is still a unique scientific answer to meet all of them: the presence of an atmosphere of neutrino particles!

To scientists, neutrino particles are still the most mysterious objects. To begin with, their existence was postulated, not discovered. Besides, scientists still have no method to directly observe and detect neutrino particles. In order for the beta-decay process to obey the law of conservation of energy, Wolfgang Pauli in 1930 postulated that this process also released an elusive particle. It was later named neutrino and was only indirectly detected 23 years later by Cowen and Reines! Now, we know that all stars produce neutrino particles during their nuclear reactions and there are at least three kinds of neutrinos. Scientists estimated that there are hundreds of millions of neutrinos passing through our bodies per second just from those produce by the Sun alone. Since there are billions and billions of stars in the universe around us, there is certainly a dense energetic atmosphere of neutrino particles in the universe. Utilizing the unique ability of the neutrinos to penetrate through matters, scientists put very large tank filled with heavy water solutions very deep underground to study neutrinos, where no other cosmic particles can reach. They have installed very sensitive light detectors in the tank to detect those high energy neutrinos, which by "rare chance" to hit molecules and are capable of breaking the chemical bonds of the heavy water or dissolved chemicals to produce very faint light. By finding the surge of energetic neutrinos simultaneously with the finding of new supernova stars, scientists have concluded that neutrinos travel at light speed. Several recent findings have confirmed that neutrinos do have very small masses.

Based on the teachings of PM scientists have long predicted the presence of an atmosphere of aether in space to produce universal forces, light, and other universal phenomena. Our scientific knowledge also confirms that the universe is only filled with an atmosphere of neutrinos, which is continuously replenished by all stars producing them at constant rates. **Since nothing else is replenished by stars, the atmosphere of neutrino particles must be the only thing occupying all space of the universe, the aether, and the common-scientific origin of all universal phenomena!** To prove that this uniquely important scientific conclusion is correct, all universal phenomena must be predicted and explainable by the presence of the atmosphere of neutrinos with their unique properties. Only the real science can meet all these difficult scientific challenges, as illustrated in this book.

B. Mechanical Gravity

Scientific Force Fields

Scientifically, force fields do exist. They are the fields of particles, which interact with other objects. A field of particles alone is a force field because that all particles have kinetic energies capable of producing forces. For example, a field of air particles produces atmospheric pressure on all objects in the field. A particle field alone does not produce a net directional force since all collision forces they produced are in random directions and are cancelled out from one another. The interactions of the field particles with other objects however often result in producing directional forces. For example, the collision interactions of atmospheric neutrino particles with all objects produce gravitational, electric, and magnetic forces. An object having energy can exchange its energy with particle fields to produce directional forces. For example, an electric fan produces wind.

For centuries scientists have postulated the existence of no-aether force fields in an attempt to explain universal forces but they still can not offer a common-sense-making mechanism for such object-free fields to produce and conduct (attraction or pushing) forces. The scientific answer is therefore the presence of an atmosphere of neutrinos to produce universal forces just like that the worldwide forces are produced by air. What a simple scientific answer, which has long been predicted by the teachings of PM! The force fields referred to in this book are therefore particle fields, which only produce on-contact pushing forces by particles.

A scientific force field is a particle field, which produces forces, such as atmospheric pressure, to act on all objects in the field. It has often been utilized to do works. For example, an electric fan produces wind in a force field of air.

Scientific (Mechanical) Gravity

1. Produced by Neutrino Particles

According to the teachings of PM gravitational force must be produced by object(s), such as particles having masses (aether). An important basic scientific principle states that when there is a (collision) interaction, there is a reaction. In other words, force, energy, or phenomena are produced by collision interactions among objects resulting in exchanging energies. When two objects, Sun and Earth for example, do not contact with each other, they can not produce any force or exchange any energy. Scientifically, gravity force needs at least two objects to have on-contact interaction to produce. Therefore, PM predicts the presence of aether in space to interact with all stars and planets to produce gravity among them, since they alone without contact can not produce gravity or any other universal forces.

According to experimental findings the strength of gravitational force is proportional to the mass contents of objects. Since all atoms and their subatomic particles such as nuclei and electrons, contain masses, aether particles need to have equal chance to collide with all of them in order to produce gravitational forces that are proportional to the mass contents of objects. It means that aether particles should be able to penetrate through objects, including their molecules and atoms to interact with all of their subatomic particles to produce gravity without damaging them. To our knowledge, only neutrinos have such unique properties! We know that they have the unique properties capable of penetrating through all objects to collide with their subatomic particles. They are so small that their collisions with subatomic particles are non-destructive. Besides, without atomic structure, neutrino particles have amazingly dense mass and, therefore, can produce such strong gravitational force already observed by astronomers. Also, they travel at light speed allowing gravitational force to have long-range effect as observed by astronomers. There are therefore very strong scientific evidences to conclude that the atmosphere of neutrinos is the aether that interact with all objects to produce gravity.

2. Possible Mechanisms

All stars produce neutrinos at essentially constant rates of their own, thus, maintaining an atmosphere of neutrino particles having a constant concentration and energy level at any giving region of the universe. Upon entering a large object, such as a star or a planet, neutrinos continuously collide with its subatomic particles resulting in continuous loss of kinetic energies until they finally get out of the object. As a result, at equilibrium, the average kinetic energy of neutrinos inside an object is lower than those outside. Since neutrinos having higher kinetic energy have stronger pushing force, there is always an imbalance in the pushing force from neutrinos resulting in having a net pushing force towards the object from all directions. For a round massive object with essentially homogeneous mass distribution there is always a force from the atmospheric neutrinos, known as gravitational force, pushing all things towards its center.

All objects on Earth are made of atoms and molecules, which are made of nuclei and electrons. Comparatively, nuclei and its orbital electrons are very small particles occupying very large volume of spaces. Scientists often say that objects are essentially empty spaces and they believe that neutrinos have very little chance to hit these subatomic particles, thus, penetrate through objects freely with essentially no collision interaction. It is very hard for them to believe that the interactions are strong enough to produce gravity.

It is possible that there are unexpected mechanisms that could enhance the interactions between neutrinos and those of the objects. For example, scientists presume that all neutrinos travel following a straight-line pass. There are however scientific evidences to show that at least some neutrinos have oscillation energies. Those having oscillation energies may not travel in a straight-line pass and have significantly more chance to collide with the subatomic particles of objects. Since neutrinos are very small and their own collisions in space are likely most sideways instead of heads-on. These collisions are expected to enhance the oscillation energies. However, since the subatomic particles of objects are much larger than neutrinos, their collisions with neutrinos are expected to reduce the oscillation energies of neutrinos. The reduction of oscillation energy of neutrinos inside objects may be an important mechanism for producing gravity. We would therefore expect that those neutrinos having oscillation energies play more important role in producing gravity than those without. Another important factor, which will be explained later, is that all subatomic particles of objects are charged particles, which have very effective way to interact with the atmospheric neutrinos.

3. The Heat Energy of Planets

Le Sage in 17th Century has proposed a gravity theory suggesting the presence of ultramundance particles taking up all space in the universe to interact with ordinary matters to produce gravity. To produce gravity the interaction must be inelastic so that ultramundance particles or neutrinos continue to lose kinetic energy, which should produce heat. According to the calculation of physicists such as Maxwell, if gravity is produced by the collision interactions of the ultramundance particles with ordinary matters, the heat produced should be so large that should instantaneously vaporize all atomic matters. For this and other reasons, Le Sage's gravity theory has been discredited and been replaced by Einstein's theory of general relativity. Physicists' calculations may be correct if the collision interactions are 100% effective to produce heat. However, the most of the energies of the interactions is consumed to produce charged particles, to keep them constantly charged, to form and to maintain atomic and molecular structures instead of producing heat. More detailed explanation will be given later.

Small amount of heat does produce from the gravity-producing interactions. Scientists have long discovered that the center of planets such as Earth is very hot. They have also observed that planets, such as Jupiter and Saturn, release more heat energy than they can receive from Sun's radiation. No-aether theories may have avoided the instantaneously vaporizing question but cannot explain the excessive heat produced by planets.

Unique Properties of Mechanical Gravity

1. No Infinitely Strong Scientific Gravity

Without scientific understanding, the mathematical models of the current gravity theories have omitted all the scientific factors affecting the strength of gravity leading to their wrong conclusions, predictions, and mathematical mistakes. According to current theories, the gravitational force between two objects is only affected by both their mass contents and the distance between them. Theory of general relativity further takes into account the shape and the mass distributions of the objects. Still, there are other important scientific factors have not been considered by current gravity theories.

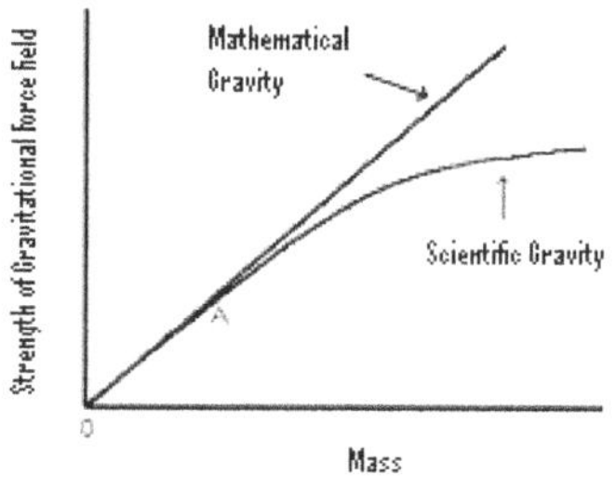

Unlike current mathematical gravity theories, scientific gravity has a maximum strength.

First of all, there is no infinitely strong mechanical gravity produced by neutrino particles. Unlike spacetime, an atmosphere of neutrino particles has limited concentration, mass density, and, thus, limited (kinetic) energy. When an object can completely stop neutrinos inside or around its body, for example, it will experience maximum pressure from the atmosphere of neutrinos, which is known as atmospheric pressure. We all know that there is a maximum atmospheric pressure of air. Similarly, when an object can absorb all the kinetic energy of the atmospheric neutrinos entering it, it gets the maximum gravity. In a thought experiment that the mass of a planet could continuously increase, its gravity would increase initially proportional to its mass. However, upon further increase in mass, the increase of gravity will begin to slow down deviating from the proportionality, when the atmospheric neutrinos can no longer reach all the particles of the object before their kinetic energy is exhausted. Finally, a maximum gravity will be reached and further increase in mass no longer has effect to the planet's gravity. It means that the black holes predicted by mathematical gravity theories do not exist. It further means that the nature does not have a way to form the "singularity" to produce the so-called big bang.

2. No Universal Gravitational Constant

Gravity should be affected by temperature, which affects the collision interaction level of particles, since the higher the temperature the faster the atoms and the more movement of their subatomic particles. Considering this scientific factor alone, stars should have much stronger gravity than planets. It means that the so-called universal gravitational constant is not applicable for star-

star and even star-planet calculations and, therefore, the mathematics of the current gravity theories is wrong for applying to stars, such as our Sun.

The universal constant is correct only for the calculation of the gravitational forces among planets! Although both Newton's and Einstein's theories have used it to correctly predict planetary motions, it is likely that the masses of all planets and Sun so calculated are off.

3. Other Scientific Factors Affecting the Gravity of Stars

The gravity of stars is much more complicated than we think and some of the reasons will be explained in more detail later in the scientific cosmology section. All stars have at their centers a "dense-matter object", which is like a giant radioactive atomic nucleus to continue their constant-rate nuclear reactions there and it has strong antigravity property by continuously producing energetic neutrinos, electron and proton. The gravity of stars is therefore not even proportional to their masses. Stars are born to 100% dense-matter objects to continue to produce hydrogen, helium, etc, to grow their outer layer having atomic and molecular matters, which produce gravity. Therefore, baby and very young stars are anti-gravity stars and their anti-gravity continues to decrease while their gravity continues to increase with aging. To make it more complicated, neutrino winds play a more important role in affecting the movements of stars and galaxies than their gravitational force. Both the misunderstanding and the mystification of the universe will get worse under no-aether postulations.

4. Space Energy

Contrarily to the current belief that the space of the universe is a vacuum having the lowest energy content, it has high energy content due to the presence of a dense energetic atmosphere of neutrinos. Neutrino particles are produced by the nuclear reactions at the centers of all stars and they have very large kinetic energy. All planets in the universe interact with the atmospheric neutrinos in such a way that neutrinos continue to lose energy inside planets and, therefore, all planets are the low-energy centers in the universe. This is apparent since everything around them falls into them. A low-energy center must have higher concentration of neutrinos than that in space since at equilibrium the number of neutrinos goes in it equals that coming out of it. Higher concentration is needed to compensate slower moving effect.

Although stars like our Sun are also low-energy centers as a whole, only their outer layers are the low energy centers. The dense-matter objects at their centers are high energy centers.

Mechanical Gravity V. Spacetime Gravity

1. Similarities and Differences

Einstein was able to predict some of the gravity-related scientific phenomena with his theory of general relativity, which had shaken the entire world. If his theory was scientifically wrong, how could he make correct scientific predictions?

According to Einstein light travels at an absolute constant speed while both time and dimension are not absolute but functions of mass distribution and traveling speed. His universe is four-dimensional spacetime distortable by environmental masses and the mass-distorted spacetime is also known as curved spacetime, which was postulated to produce gravitational force.

Spacetime alone in the absence of massive stars and planets is not distorted and remains unchanged. In the presence of a massive body spacetime is so distorted that it curves around the massive body. According to Einstein, the curvature of spacetime causes time to slow down upon approaching the massive body. The difference in time between a location closer to and a location farther away from the massive body causes things to fall into the massive body. Mass-distorted or curved spacetime is therefore the scientific origin of gravity. The curvature of the spacetime increases upon approaching the massive body. Inside it, the curvature of spacetime begins to decrease and reduces to zero at its center.

Although theory of general relativity is no-aether postulations based, assuming something (spacetime) in space that can be affected (distorted) by matters happens to mimic having-aether science.

There therefore some amazing similarities between the science of PM and the theory of general relativity. For example, similarity exists between the curvature-variation of spacetime and the energy-variation of the atmospheric aether (neutrinos). Like spacetime, the energy level of the atmospheric neutrinos is unchanged in the absence of a massive body. It also changes around a massive object similar to the distortion of spacetime due to their interactions with the massive body. Upon approaching the same massive body, the kinetic energy of the atmospheric neutrinos decreases. Inside the massive body, the kinetic energy of the atmospheric neutrinos continues to degrease and reaches a minimum level at its center. This similarity is shown in following picture.

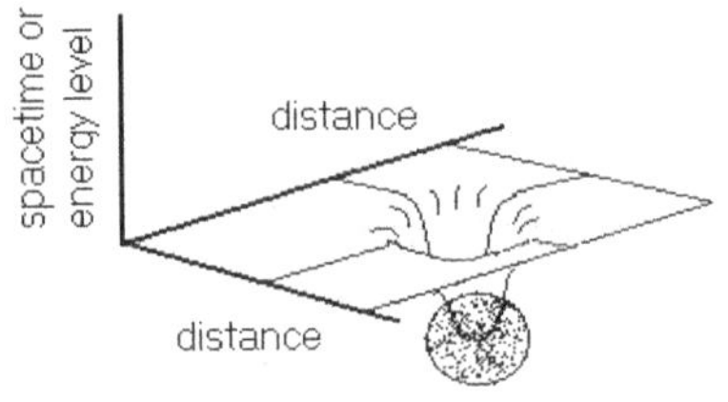

The graph shows the amazing similarity of the variation of the curvature of spacetime to that of the (kinetic) energy of the atmospheric neutrino particles, caused by their interactions with a massive body.

These two versions of interpretation of gravity are however scientifically incompatible to each other and one of them must be wrong. Spacetime is not an object having mass but a mathematical model based on postulations that both time and dimension are not absolute, which are further based on the postulations that light travels at an absolute constant speed in space without a transmission medium. If theory of general relativity were correct, its mathematical model of spacetime should be able to interpret all forces such as electric force, the atmospheric pressure of air, etc but it can not.

All no-aether theories were developed due to the longtime failure of finding aether. Many physics text books still state the fact that Einstein's theory of relativity or modern physics has been accepted because some findings have proved that it is correct. They further state that aether can not be found and there has no scientific evidence to prove the presence of aether. What these text books are saying is that once aether was found, theory of relativity would be disproved. Now, aether has been found and the fundamental no-aether postulations of the theory of relativity have been proven both wrong and obsolete. The correct interpretation of gravity is the mechanical gravity given above that is consistent with the interpretation of all forces.

2. Light Bending and Gravitational Red-Shift Prediction

Einstein's most famous prediction was that even light was bent by the curved spacetime and it was proven correct. The new aether science also predicts that light pass should be bent towards the lower kinetic energy direction of the atmospheric neutrinos. There are now undeniable amounts of experimental findings supporting the presence of a dense atmosphere of neutrino particles to produce gravity. To support Einstein's theory of general relativity, physicists also need to have large amounts of experimental findings to prove that curved spacetime does exist and that the atmosphere of neutrino particles either does not exist or is not responsible for producing gravity.

Einstein further predicted that strong gravity would cause light to red-shift. Although the gravitational red-shift caused by the gravity of the Earth is generally too small to be detectable, physicists may have detected the red-shift of F_e^{57} in their Mössbeauer Effect experiments. They have also observed the red-shift of sodium light from the Sun, etc. Again, the new aether science also predicts similar "red-shift" phenomena from the kinetic energy difference of neutrino particles. There is however subtle difference between the two predictions and may be differentiated by experimental findings since the gravitational red-shift of general relativity is directional but that of new aether science is not. For example, according to the new aether science a spectrum taken in an underground laboratory should show red-shift in comparison to that taken in a space station. This non-directional red-shift Doppler Effect of light is not expected by theory of general relativity but by the new aether science.

3. Time Difference v. Energy Variation

According to theory of general relativity curved spacetime has time difference among different locations, which is the scientific reason for having both gravitational red-shift and light bending effects. There are two time effects: 1. increasing moving speed slows down time; 2. time slows down upon approaching a massive body. Physicists have long been doing experiments to detect time difference in an effort to scientifically prove theories of relativity. According to new aether science the energy variation of the atmospheric neutrinos should affect the rates of nuclear reactions such as nuclear decay rates, thus, affect atomic-clock time. New aether science therefore predicts that the atomic clock gets faster upon approaching a planet, which is exactly opposite to Einstein's prediction! If travel speed has no effect on atomic-clock time, all the experimental findings in the past favors the author's prediction that atomic-clock time goes slower in space than on Earth.

New aether science continues to support that time is absolute and there is no difference in time between any two locations. There is however no method to very accurately measure time, particularly in different locations having differences in kinetic energy and concentration of atmospheric neutrinos. Atomic clock is no longer valid for measuring absolute time in this situation.

There have been two publicized experimental findings in time difference studies. One showed that the atomic clock on-board a fast-moving satellite went slightly slower than a synchronized atomic clock on Earth. The other was the need to correct the "atomic-clock time difference" in the Global Positioning System (GPS) to make GPS to work correctly. Both findings however supported the prediction of the new aether science that the atomic-clock time accelerates upon approaching Earth. These results are however inconclusive for theory of general relativity since they have not yet independently proved any of its two predicted time effects.

Drag Effect and Planetary Motions

Having the ability of penetrating through matters essentially freely, the drag effect of the atmospheric neutrinos is small for slow moving objects but it should slow down all moving object anyway. The drag effect should be the reason why all satellites and space stations only stay in orbit for a few years and eventually fall back to Earth and for spaceships to slow down in space. The drag force of the atmospheric neutrinos however can be amazingly strong for very fast moving objects, for example, for particles moving close to light speed needing to encounter the atmospheric neutrinos moving at twice the light speed. It is therefore not surprising that physicists can not accelerate electron or any other kinds of particles to light speed. Again, they religiously believe the reason is the effect of the theory of special relativity that mass increases with speed instead of simply the drag effect of the aether (atmospheric neutrinos).

One of the biggest secrets of the nature is that planets and their moons in the solar system can circulate Sun essentially forever without slowing down and falling into the Sun. It gives the scientists the strongest reason to believe that there is no aether in space, since the drag effect of aether should slow planets and moons down. Of course, it is also the most often raised question by physicists against the aether concept of PM and this new aether science. Actually, it is one of the most difficult scientific questions for scientists to answer with or without aether. Although it is used to support no-aether theories, the solar system is just too orderly for scientists to conclude that it occurs naturally unless there is god to design it. Today's physicists support no-aether theories but they can not explain such orderliness as that all nine planets orbit the Sun in the same direction and essentially on the same plain.

In the presence of atmospheric neutrinos, for planets to stay in their orbits, they need constant energy supply to overcome both the drag effect of the atmospheric neutrino particles and Sun's gravitational force. The only scientific possibility is that the Sun constantly supplies the force needed by all 9 planets and their moons to stay in their orbits. The new aether science offers such a possibility.

It will explain in Scientific Cosmology that there is a dense-matter object inside all regular stars producing their nuclear reactions. The so-called neutron stars are actually exposed dense-matter objects by supernova explosions of stars. Having a very high-energy surface layer constantly undergoing nuclear reactions, all dense-matter objects inside regular stars have antigravity properties. Like many neutron stars the dense-matter objects inside many stars such as our Sun have fast rotating dense-matter objects with very strong electric and magnetic force fields around it. The dense-matter object inside a star has a large space separating it from its outer layer and therefore the fast rotation of its dense-matter object does not affect its visual shape. The outer layer of a regular (mature) star has strong gravity, which has overcome the antigravity of its dense-matter object to become a gravity star.

The antigravity effects of the dense-matter objects in some stars may not be completely overcome in all directions. The high-energy neutrino particles produced by the dense-matter object of the Sun affected by its rapid rotation and its strong electric and magnetic force fields may form a constant spiral neutrino wind that provides a directional pushing effect only. The existing of such a spiral neutrino wind may cause (the outer layer of the) Sun to rotate and may be utilized by all planets to stay in steady-state orbits! If a planet is caught by the Sun in the right direction, such a spiral wind at equilibrium would constantly give it a push in both its orbiting and anti-gravity directions to keep it in a steady-state orbit. Such a spiral wind of neutrinos should have much stronger pushing force near the surface of the Sun and get weaker quickly outward. Planets closer to the Sun, therefore, orbit faster and need stronger push to overcome both the gravitational force of the Sun and the drag effect of the atmospheric neutrinos to stay in orbit than farther away planets. Of course, all planets must orbit in the same direction to survive.

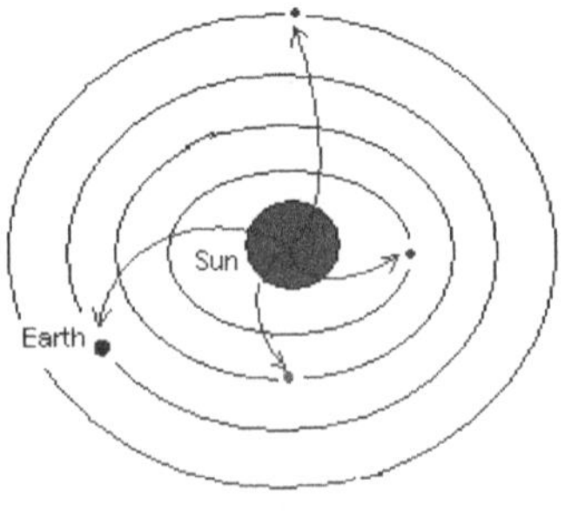

Our Sun may produce a permanent spiral wind of neutrinos to cause its outer layer rotating and to keep its planets in steady-state orbits.

Since all the above given difficult nature conditions are met by all nine planets of the Sun, this explanation has very strong scientific evidences to support it and becomes plausible. Once the Sun's gravity catches a new planet, it will take a long time for it and the rest of the planets to reach new equilibrium orbits of their own. Finding no planet orbiting Sun in clockwise direction suggests that those planets have not survived. Statistically, there may be around equal number or nine planets captured by Sun circulating clockwise having plunged into Sun!

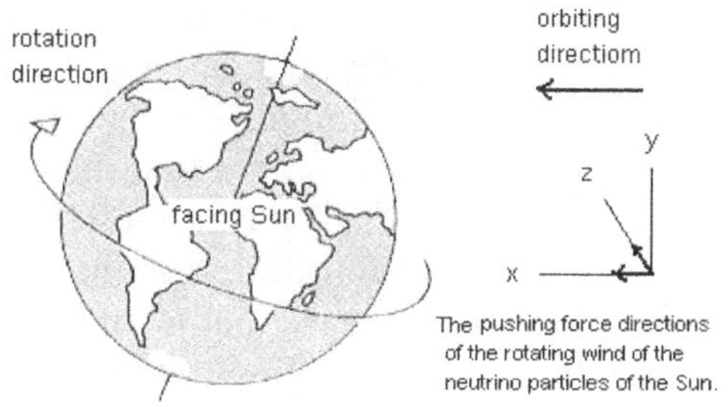

> The spiral wind of neutrino particles of the Sun has a pushing force on Earth in both directions, away from Sun (z-force) to counter gravity and in its orbiting direction (x-force) to keep Earth from slowing down.

It is also reasonable to conclude that the same spiral wind of neutrinos keeps the moons of planets in orbits too. Most moons also orbit in the same counterclockwise direction with a few exceptions, which are the outer moons of Jupiter: Anake, Carme, Pasiphae and Sinope; of Saturn: Phoebe; of Neptune: Triton. Such exceptions may be allowed to reach a minimum-energy state at equilibrium for planets having multiple moons. It is however of interest to study whether these moon really have steady-state orbits.

In summary, the force of the spiral wind of neutrinos of the Sun should have two components, the x-force in the planet orbiting direction to counter the drag of the atmospheric neutrinos and the anti-gravity force pushing planets away from Sun or the z-force. Since Earth is in a steady-state orbit, we do no feel any force acting on it.

A satellite or space station orbiting in the direction the Earth and the Moon orbit should have longer lifetime than those orbiting in the opposite directions. In theory, those having maximized its utilization of the spiral neutrino wind of the Sun should be able to stay in orbit essentially forever like what planets and their moons do.

The same spiral neutrino wind of the Sun should also provide both the Sun and the planets with rotational energy. One possible mechanism for planets is that the spiral wind has stronger pushing force on the half of the planets closer to the Sun than on the half farther away. If this is the correct mechanism, the larger a planet, the larger the difference in the pushing forces on both halves and the faster a planet should rotate. It means that giant planets Jupiter and Saturn should be the fastest rotating planets in the solar system and they are!

C. Other Universal Forces and Energies

Like the presence of an atmosphere of air, the presence of an atmosphere of neutrino particles should produce a lot more universal phenomena besides gravity. Since neutrinos are the only particles filling the space of this universe continuously replenished by stars, they should be responsible for producing all universal phenomena including all universal forces and light. Besides gravity, scientists have discovered other universal forces including strong nuclear force, weak (nuclear) force, electric force and magnetic force. They all should be produced by the interactions between the neutrinos and ordinary matters and explainable by the unique properties of neutrinos. Sure universal forces are expected and they all have the unique physical properties of neutrinos! Besides, the atmospheric neutrinos should also be responsible for producing all universal energies too.

Here is a good example about the difference between correct science and wrong science. Einstein's dream was to expand his theory of relativity to all universal forces to develop a theory of everything. This has become a common dream for physicists too. Their dream however can never come true because they have started with wrong or unscientific postulations. Only the correct science is itself the science of everything.

Strong Nuclear Force

Modern physics - The nuclei of atoms are made of proton and neutron particles. Since protons are positively charged particles, scientists believe that there exists very strong force to "glue" them together to overcome the repulsion of the positively charged protons to form nuclei and called this universal force the "strong (nuclear) force". Their thinking is that a force which can hold a nucleus together against the enormous forces of repulsion of the protons is very strong indeed. However, they found this force having very short range of effectiveness unlike the "inverse square force" of the electromagnetic force and gravity, which have very long range. A Japanese physicist Yukawa modeled the strong nuclear force as an "exchange" force in which the exchange particles are pions and other heavier particles. According to standard model of particle physics, the range of a particle exchange force is limited by the uncertainty principle. Protons and neutrons, which make up the nucleus, are themselves considered to be made up of quarks, and the quarks are considered to be held together by the color force.

New aether science - All universal forces are produced by the atmospheric neutrino particles. We have learned that an atmosphere of air produces atmospheric pressure on everything inside its atmosphere. However, we do not expect that the atmosphere of neutrino particles has an atmospheric pressure on all objects having atomic and molecular structures since neutrinos penetrate right through them. The only thing they may not be able to penetrate through, at least not easily, may be the nuclei of atoms, which has amazingly dense masses. It means that there should be a full or partial strength atmospheric pressure of neutrinos on all the atomic nuclei. This is a much stronger force than gravitational force since it is a direct pressure of all neutrino particles while gravitational force is the differential force between the neutrinos entering objects and those coming out of them.

Unlike gravity, electric, and magnetic forces, strong nuclear force was found to be very short ranged and therefore has the unique properties of an atmospheric pressure, which become ineffective in the presence of very small space or hole allowing atmospheric neutrinos to enter or to pass through and is therefore very short ranged indeed.

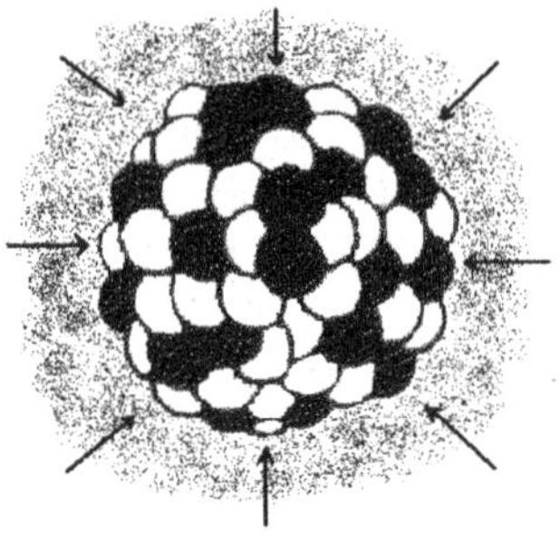

The atmospheric pressure of neutrinos is the scientific origin of the strong nuclear force, which keeps the nuclei of atoms stable inside the universe.

Weak Force

Scientists have found the existence of weak (nuclear) force, which causes radioactive decays such as beta decay.

Modern physics – According to the standard model the weak interaction involves the exchange of the intermediate vector bosons, the force particles W and Z. Since the mass of these particles is in the order of 80 GeV, the uncertainty principle dictates a range of about 10^{-18} meters which is about 0.1% of the diameter of a proton. The weak interaction is the only process in which a quark can change to another quark, or a lepton to another lepton - the so-called "flavor changes".

New aether science - The weak force again is an expected universal force simply due to the presence of an energetic dense atmosphere of neutrinos. The neutrinos produced by the nuclear reactions of the stars are very energetic and their constant bombardments cause most cosmic particles and even some atomic nuclei to decay.

Charged Particles, Electric and Magnetic Forces

There are conclusive amounts of experimental results to prove that two kinds of charged particles exist, the so-called positively charged and the negatively charged. Particles carrying the same kind of charges repulse each other but those having different kinds of charges attract each other. Unlike gravitational force acting on everything, electric and magnetic forces are selective forces, which only act on certain kinds of matters - those carrying charges and those having lone-pair electrons, respectively.

Ordinary matters such as those found on Earth are made of atoms and molecules which do not carry any charge. However, atoms and molecules are made of two fundamental particles only: proton and electron. Both are charged; Proton carries positive charge while electron is negatively charged.

1. Currently accepted Concepts - Like gravitational force, both electric and magnetic forces are explained using both the force field concept of Faraday and modern physics. Electric force and magnetic force have been found to have the same scientific origin and given a combined name: electromagnetic force, which has infinite range obeying the inverse-square law, thus, having the same mathematical form as the gravity force. It has been further modified to obey both theory of relativity and quantum theory and is now called the theory of quantum electrodynamics or QED. According to QED, electromagnetic force is the universal force involving exchanging photons.

2. New aether science - Both electric and magnetic forces should be produced by neutrino particles like other universal forces from their collision interactions with the subatomic particles of ordinary matters.

a. The Composition of a Charge Particle

Charge should be some kind of energy and a charge particle constantly carries this "charge" energy. To do so, charge particles must have a constant energy supply from the environment. Since the universe is filled only with an atmosphere of neutrinos, charge particles must obtain their constant energy supply from the atmospheric neutrinos. By doing so, a charge particle becomes a low energy center in the atmosphere of neutrinos similar to a planet. Like a planet, a charge particle should have gravitational force and its own atmosphere of neutrinos, which should be denser and have lower (kinetic) energy than the atmospheric neutrinos of the universe. Comparing both the size and the amount of charge energy they carry, an electron should have much denser atmosphere of neutrino particles and much stronger gravity than a proton.

A charge particle therefore is made of an absolutely uncharged particle (AUP) having an atmosphere of neutrinos of its own. A charge particle must have an AUP having some kind of unique ability to effectively and continuously getting the energy of neutrinos to become constantly charged. The detectable properties of charge particles appear to be the common properties of their atmospheres, not those of the AUP at their centers! As a result, it appears difficult to study how an AUP, of electron for example, can continue to consume the energy of neutrinos. What we have been finding are the charge properties of protons and electrons, which are those of their atmospheres of neutrinos, not their AUPs' hiding inside.

Both proton and electron are charged, thus, have their own atmosphere of neutrinos, which make up all ordinary matters, for example, everything on Earth. Neutron is only a proton-electron pair. Neutrino particles inside ordinary matters such as Earth are therefore unevenly distributed. There are free neutrinos having higher kinetic energy and neutrinos inside the atmospheres of protons and electrons having lower energy. At equilibrium, the same amount of free neutrinos enters and comes out of their atmospheres in exchanging of energy needed to maintain their charge property or energy. The energy of neutrinos inside the atmospheres of both proton and electron reduces upon approaching their AUP while their concentration increases. The neutrinos deep inside the atmospheres of charged particles, particularly electrons, may have acquired some kind of orderliness, such as liquid-like.

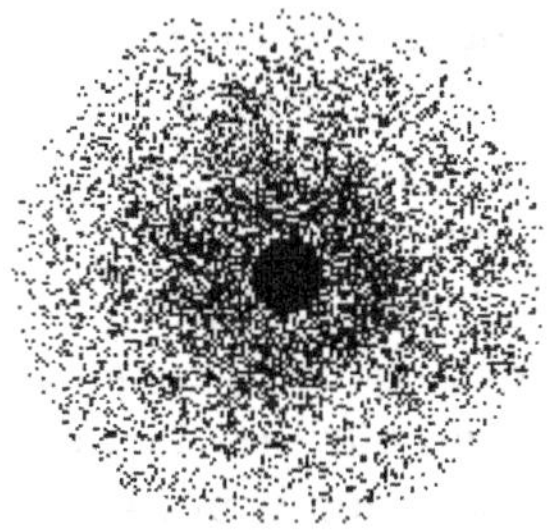

Both proton and electron have their own atmosphere of neutrino particles and gravitational force.

The universe has only one thing filling all space that is constantly replenished by all stars: the atmosphere of neutrinos. There therefore can have only one kind of charge particles: AUP having an atmosphere of neutrinos. In other words, there can not be two different kinds charged particles and, therefore, can not have anti-matter.

b. The proton-electron universe

A possible scientific reason why charge particles of the same kind repulse each other is that their own bulky atmospheres of neutrino particles push each other away. But why a positively charged particle attracts a negatively charged? The major difference between the two elemental charge particles is their large size difference. Proton is much larger than electron. In this case the gravitational "pull" between the two particles may have overcome the repulsion force between their atmospheres of neutrinos resulting in "attracting" to each other. Why electric force has no effect on non-charged matters? Probably because that all non-charged matters are made of equal numbers of positively charged and negatively charged particles and the electric force acting on them simply cancel out because of that. This reasoning applies to all positively and negatively charged particles made of protons and electrons of all sizes since atmospheric neutrinos simply ignore the neutralized part of a particle. For example a Ca^{2+} attracts to a CO_3^{2-} despite that the size difference of these two charged ions are much smaller than that between a proton and an electron. Since atoms and molecules are made of charged particles and electric force plays a dominate role in making them, understanding the presence of an atmosphere of neutrino particles is of fundamental importance in science.

Since the concentration of the atmospheric neutrinos of a charge particle increases upon approaching its core or AUP, a charge particle eventually will repulse an approaching charge particle. As a result, inside atoms and molecules Coulomb's Law does not apply since electrons do not fall right into protons and positively charged nuclei as expected by the positive/negative charge concept. Instead, at very close distance a proton and an electron begin to repulse each other. Therefore, electrons orbit around proton and atomic nuclei instead of falling into them. The bulky atmosphere of neutrinos producing repulsion force among electrons would force themselves taking up quantized orbital. For a charge particle and an atom to continue to function, they require continuous energy supply from the neutrino particles of the universe, again, like a planet to maintain its gravity field. The atmospheric neutrinos of the universe therefore play a vitally important role in continuing supply energy to keep charge particles charged and their atoms and molecules stable.

Experimental findings show that it is not easy for an electron and a proton to form a neutron. Instead, it is more stable for them to form a hydrogen molecule. However, two protons and one electron would form a stable nucleus with an orbiting electron. In this nucleus system, the electron inside the nucleus appears to play an important role in connecting two protons at very close distance probably by mutual gravitational force and this nuclear structure requires an orbital electron to neutralize the whole system to stabilize. It is unlikely that there is a clear cut of having a neutron and a proton in this nucleus. Clearly, heavier and stable atomic nuclei are formed under the same ratio of proton : electron = 2:1 and the right amounts orbital electrons to have a balanced charge of zero.

Experimental findings reveal that free electrons and protons are very slightly heavier than those inside atomic nuclei. Also, the protons and the electrons in heavier atomic nuclei are very slightly heavier than those of lighter atomic nuclei. It means that both protons and electrons need to release some of their atmospheric neutrinos to form atoms and to from heavier atoms in nuclear fusion reactions, known as loss in masses. Since in a nuclear reaction the total number of protons, neutrons, and electrons are accounted for, the large amount of energies released in nuclear fusion reactions must be due to their sudden release of some of their neutrinos. The mechanism may be similar to rapid evaporation of liquid air resulting in an explosion.

The AUPs of both proton and electron are so small that they should have only very slight chance to collide with a neutrino particle to exchange energy. However, having an atmosphere of neutrinos of their own, they are able to effectively interact with free neutrinos to constantly exchange energy. Having much denser atmosphere of neutrinos electron is particularly efficient in interacting with the atmospheric neutrinos of the universe. This is evident from the fact that when an electric current starts to flow, it instantly creates electric force. It is also likely to be the major reason why the atmospheric neutrinos of the universe can only and very effectively exchange the oscillation energy of electrons and carry (transmit) them through the space of the universe as lights.

Inside ordinary matters the atmospheric neutrinos of the universe interact with the atmospheric neutrinos of both protons and electrons to effectively exchange energy keeping both proton and electron constantly charged and their atoms and molecules stable. This is the major mechanism of the collision interactions between the atmospheric neutrinos of the universe and the atmospheric neutrinos of protons and electrons of the ordinary matters that results in producing gravity. Only a small fraction of the energy lost by the neutrinos of the universe is used to induce the atomic motions, such as the oscillation, of electrons and protons, which produce heat. The general belief among physicists that to produce gravity the heat generated by the collision interactions between neutrinos and ordinary matters would instantly evaporate all matters; is therefore scientifically wrong!

There is one kind of charge particle made with an AUP having an atmosphere of neutrinos of its own. It may be true that the same charge particles repulse one another but it is not always true that different charge particles attract one another. Therefore, our knowledge of the electric and magnetic properties is limited or applicable only to the proton-electron system, which is unique to matters made of protons and electrons. For example, it is not applicable to the study of those charge particles found in cosmic rays and produced in high-energy laboratories!

c. Mechanical Energy of Neutrinos

The concept of having nothing (no aether) in space has long been accepted and therefore it has been considered impossible for a spaceship to use a mechanical engine. However, under an atmosphere of neutrinos it becomes possible to utilize the mechanical energy of neutrinos to do so. After all, the mechanical energy of neutrinos has been used to produce all universal phenomena! Although neutrinos penetrate through all matters having atomic and molecular structures, they do interact with charge particles to produce force. Hopefully, technological breakthrough in space science and other scientific fields will shortly occur upon recognition of neutrinos to be the aether scientists have looked for centuries.

Heat and Temperature

These topics are part of the physics of mechanics, which are well understood science. Heat phenomena can be interpreted in terms of atomic and/or molecular motions and temperature is the measure of the kinetic energy of their motions. Ideal gases have non-polar atoms and/or molecules such as H_e and H_2. In terms of absolute temperature scale, the volume of an ideal gas reduces to zero under any pressure at 0 ^oK. Scientists have found that at very close to absolute zero temperature, atomic structures are no longer stable since their orbiting electrons fall into nuclei to form extremely dense matters, which are called "dense-matter objects" by the author. The findings suggest that matters having both atomic and molecular structures do not exist in subzero ^oK temperature environment. In the absence of atomic and molecular matters, life is no longer meaningful. However, the existence of the subzero ^oK environment is predictable and is important scientifically. To measure the energy levels of this environment, the measurement of the kinetic energy level of atmospheric neutrinos is needed.

Scientists have long been wondering what happens to the predicted huge amount of heat that should produce by the collision interactions between neutrinos (aether or ultramundance particles) and ordinary matters to produce gravity. The scientific answer is given here. Free neutrinos constantly enter and come out of the atmospheres of neutrinos of the electrons and the protons in atoms and molecules. This energy exchange mechanism among neutrinos themselves is not effective in causing atomic and molecular motions and is needed mainly to maintain the charge properties of protons and electrons and their atomic and molecular structures. However, the constant bombardment of neutrinos on the neutrino atmospheres of protons and electrons and their AUPs may cause them to oscillate since having very small sizes they should not be evenly bombarded by incoming neutrinos from all sides. This is similar to the Browning Motion phenomenon scientists have long observed. The oscillation motion of electrons and protons should exist at all temperatures and constantly produce heat. The amount of heat produced should be proportional to the environmental mass density. It is well known that heat propagates through solids, liquids, and gases but is rather slow. When the heat continuously produced by the environmental matters is faster than the overall heat transferring process, heat accumulates by rising temperature or producing faster and more atomic and molecular motions until heat transfer rate catches up the heat production rate to reach an equilibrium temperature. This explains why the centers of planets have the highest equilibrium temperature and why all planets release more heat than the heat they absorb from the Sun.

Winds of Atmospheric Neutrino Particles

Since all stars continuously produce neutrinos, each galaxy should produce a galaxy wind constantly blowing out in all directions. There are lots of galaxies so are their winds. When their winds encounter one another, a lot of spiral or rotational winds of neutrinos, like hurricanes on Earth, are routinely present in the universe.

It will be explained in more detail in the Scientific Cosmology section that all stars have a dense-matter object at their center, which is made of protons and electrons similar to the nuclei of atoms but is much bigger. Like radioactive nuclei a dense-matter object constantly releases protons, electrons, neutrinos, and huge amount of energy to start and maintain nuclear reactions on its surface. Also like atomic nuclei, a dense-matter object has very dense mass and even neutrinos may not be able to penetrate through it. As a result, the force of neutrino wind can effectively act on dense-matter object or stars. Neutrino winds therefore play a major role in affecting the movements of stars and galaxies particularly on very young stars and galaxies, whose stars have very high dense-matter object contents. This is evident from the findings that there are many spiral galaxies, which are rotating. Astronomers also found that the movement of stars in a spiral galaxy does not fallow the rules of the movement of the planets in the solar system, which is believed to be a gravity-maintained system. For example, the stars closer to the center of the galaxy move at about the same velocity as those farther away from the center. These findings are strong scientific evidences to show that the movement of stars in a galaxy is mainly affected by the wind of neutrinos, not by gravitational force as believed by today's scientists.

There are many spiral galaxies explainable by that the movements of galaxies are affected by the spiral winds of neutrinos.

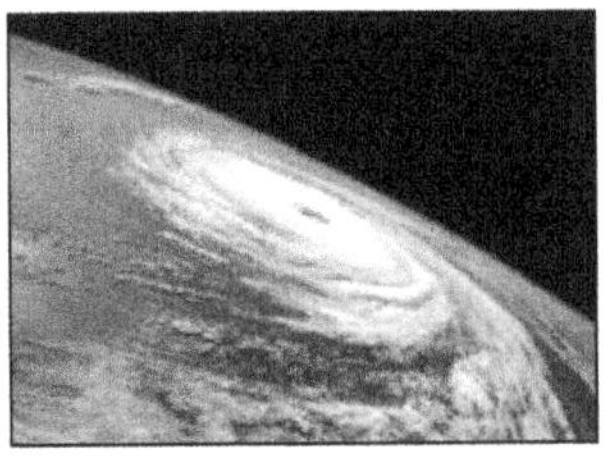

A hurricane is shaped by spiral wind so may be many galaxies.

A Revisit of Lesage Gravity

Georges-Louis Le Sage in 17th Century proposed the presence of a sea of (ultramundance) particles, which like today's neutrinos were postulated capable of penetrating through all matters and gravity was produced by their constant bombardment of ordinary matters. Serious doubts as given below have been raised on this theory and physicists have long abandoned it to accept theory of general relativity. The mechanical gravity of the new aether science is essentially the same as that of Le Sage except that neutrinos have been identified to be the postulated ultramundance particles. The following serious scientific problems encountered by Le Sage theory therefore must be satisfactorily answered to bring scientists back to mechanical gravity.

1. The collision interaction of neutrinos with ordinary matter must be inelastic collisions in order to produce gravity. According to the calculations of some physicists such as that of Maxwell, the heat produced by such interactions would evaporate matters in seconds!

This question was raised because of the lack of scientific understanding of the interactions between neutrinos and ordinary matters. Energy transfer must happen among the collision interactions to produce gravity. If the interactions are for the ordinary matter to produce 100% heat, physicists' calculation and prediction may be correct. However, the collision interactions are mainly to produce charged particles, to maintain their charge energy, atomic and molecular structures, instead of producing heat. A charged particle is made with an AUP having an atmosphere of neutrinos of its own. To be constantly charged, it continues to absorb the energy from the "free" neutrinos of the universe. When a neutrino enter an object, they hit the atmospheres of neutrinos of the electrons and the protons to lose energy and this process involving neutrino-neutrino collision interactions that essentially does not produce heat! The constant bombardments of neutrinos on electrons and protons do however make the entire atom to vibrate and their vibration is the scientific origin of heat or temperature. Therefore, small amount of heat is produced by the overall collision interactions between neutrinos and ordinary matters. This can explain why the planets in the solar system release more heat than that they can absorb from the Sun and why their centers are hot. An environment reaches a steady-state temperature depends on both the kinetic energy level of the atmospheric neutrinos and the mass content of the ordinary matter that has rather slow rate of heat transferring processes.

2. According to the calculation of Pierre-Simon Laplace the speed of gravitational force in Le Sage's theory is "at least a hundred millions of times greater than that of light". At lower speeds aberration would cause the apparent force on the orbiting bodies to point slightly behind the current position of the source, causing the

bodies to recede from each other. This "aberration" question still bothers many physicists.

There is a serious question on the science of Pierre-Simon Laplace's calculation. Why gravitational force has propagating speed? If there is one, it should be around the speed of the atmospheric neutrinos or the speed of light. A heavenly body has a steady-state gravitational field around it and therefore, its gravitational action is instantaneous without any "aberration" effect. No aberration effect has been found anyway.

3. Scientists believe that there is no frictional drag on planetary motions.

There are conclusive proofs that the rotational motion of the Sun, the orbiting and the rotational motions of planets and moons in the solar system are steady-state motions. It appears impossible for any aether theory to explain the existence of such motions. However, upon scientific understanding the universal phenomena and the universe, a scientific explanation to these steady-state motions has been found as described in the above Planetary Motions section. The answer is that there is frictional drag effect by the atmospheric neutrinos on all moving objects. However, the Sun produces a spiral wind of neutrinos to produce and to maintain all the steady-state motions in the solar system.

D. Mechanical Light and its Universe

Misconceptions

Like gravity, light is one of the most studied subjects but the least understood. Worse yet, it has long been the most controversial and most misunderstood object. Having so many misconceptions about both light and gravity that in the past century Einstein and physicists have developed a religious pseudo-science based on them. A major re-education is needed to correct these misconceptions. Let's get started right here.

1. Not a Mathematical Equation

Light is a phenomenon that must be produced by an object. It can not be electromagnetic wave, which is a mathematical equation describing the postulated electric and magnetic force fields postulated to exist by Faraday but still unproved. Yes, the speed of electromagnetic wave happens to be the same as that of light. It is just a coincident, since physicists have neither proved that electromagnetic wave has any scientific meaning nor that it is light. Hertz in 1888 discovered an invisible light known as radio wave produced by fast oscillating electrons. He however had neither proved that electromagnetic wave was light nor discovered what light was.

2. Not Pure Energy

Only objects having masses and relative motions can have and can produce energy. Object-free pure energy has no scientific meaning and does not exist. Light is not an object-free pure energy. It means that light must have a transmission medium.

3. Having Aether to Transmission Light!

There are undeniably large amounts of experimental findings proving that light has transmission medium. For example, light has all the unique properties of the particle wave of the sound. Light also has all kinds of Doppler Effect the sound has. The science of sound has long been well understood and should be utilized to scientifically understand light. Having such strong similarity in so many physical properties, the science of light should be very similar to that of sound. Like sound, light should have a transmission medium. Also, according to the teachings of PM, light is a phenomenon having energies, it must be produced by an object (aether). Now, aether, the medium of light long predicted by scientists, has been discovered to be the atmospheric neutrinos.

4. Not a Transversal Wave

Instead, light is the particle wave of neutrino particles like that sound is the particle wave of air.

5. No Absolute Constant Speed

For any object or phenomena to have absolute constant speed, it has no relative motion to all other objects, thus violating the fundamental physics law that all objects have relative motions. Scientists have not found anything else traveling at an absolute constant speed. Yet many of them religiously believe that light travels at an absolute constant speed in space. Even today, physicists still rely on the negative findings of the over-a-century-old Michelson-Morley experiment as the proof of Einstein's postulations that there is no aether in space to transmit light and that light travels at an absolute constant speed. This experiment is based on the reasoning that since Earth constantly has both rotational and orbiting motions, on its surface light coming and going in different directions should travel at different velocities. Michelson-Morley experiment used light interference method to detect this expected micro difference in light speed! The experiment split the same light beam into two in different directions and then recombined them. Physicists believed that having different velocity would have changed their wavelengths and the recombined light beams should have light interference to each other. The results were negative and physicists had concluded that no aether existed in space and that the finding supported Einstein's postulations that light traveled at an absolute constant speed in space by itself.

However, the so-called light interference through double slits is still a very poorly understood science and the application of light interference to detect the micro difference in light speed has not yet been proven valid. The fact that we can see everything so clearly proves that lights of different wavelengths do not inter each other. In 1992 NASA's COBE satellite study of the cosmic microwave background found by A. Penzia and R. Wilson in 1963 happened to use the same scientific principle of the Michelson-Morley experiment. This time NASA scientists intended to prove the inverse that microwave background should be homogeneous in all directions having no difference in light speed. Having a homogeneous residue cosmic microwave light background is also predicted by big bang theory and many physicists are very interested in getting some experimental findings to support this theory. This experiment however used a different method to determine light-speed difference. It directly measures the wavelengths of the microwave-background light coming from different directions instead of using the light-interference method used by Michelson-Morley experiment. It surprisingly found dipole anisotropy property in the incoming cosmic microwave background light, which strongly proved the presence of aether and disproved the validity of the Michelson-Morley experiment, which apparently had an invalid detection method.

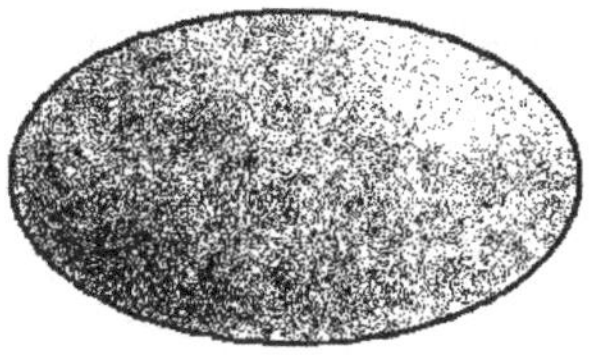

Cosmic microwave background lights from different directions travel at different velocity towards Earth.

What NASA had found was that the wavelength of the incoming cosmic microwave background light from the southwest direction was the shortest (darker color) indicating the light velocity is the fastest. Meanwhile, the wavelength from the northeast direction is the longest (lighter color) or the light velocity is the slowest. The change in wavelength is gradual from the southwest to northeast making this dipole anisotropy property very difficult to be explained by any other scientific reasons. NASA therefore had found aether and that Earth was moving at an "absolute velocity" in it. This strong positive result, therefore, had disproved the validity of the Michelson-Morley experiment and its conclusions. In other words, NASA has disproved the fundamental postulations of theories of relativity that there is no aether in space to transmit light and that light travels at an absolute constant speed in space! The uniqueness of the cosmic microwave background will be further discussed in the Energy Sources and Backgrounds of Light and Neutrino Winds and Light sections.

The entire physical science society however has ignored the important meaning of this finding! Instead, they have made an artificial "correction to the dipole anisotropy phenomena" to obtain a homogeneous cosmic microwave background and advertised this altered data to be the experimental proof of big bang theory! What a joke on science! The above findings and conclusions however have been pointed out by a few physicists and their comments are given below.

From the "Little Book of the Big Bang" by Craig J. Hogan published in 1998: "Maps of the cosmic background radiation tell our velocity through the universe quite precisely – both how fast we are going and in what direction – to an accuracy of a few percent, better than a car speedometer. It is significant that there is such a thing as a velocity through the universe, because absolute velocity is not defined by the laws of (modern) physics per se, which recognized only relative velocity; indeed, this "principle of relativity" is a cornerstone of our theory of spacetime. However, the distribution of the matter and radiation of the universe does define a "preferred frame" that, at each position, in the frame in which the universe looks the same in all directions. The solar system is moving relative to that frame, with a velocity of 375 kilometers per second, toward the head of the constellation Virgo." Also, in the book "Ripples in the Cosmo" by Michael Rowan-Robinson published in 1993: "When allowance was made for the Earth's motion around the Sun and the Sun's motion around the galaxy, the net result was that our Galaxy was moving through space at a speed of 600 kilometers per second. This seemed an alarmingly high speed. There is a curious aspect to this. The famous Michelson-Morley experiment of 1985 showed that the speed of light in a vacuum is the same whatever the speed of the observer. This destroyed the concept of the aether, which had dominated nineteenth century physics, and launched the way for the special theory of relativity in which all observers in uniform motion with respect to each other are equivalent. In cosmology, however, the situation is different. There appears to be an absolute cosmological frame of rest in which the universe looks perfectly isotropic. We are not in that frame so we see the microwave background dipole. In writing about the discovery of the dipole for *nature* in 1977, I ironically headlined my piece "aether drift detected at last."

If there is an "absolute cosmological frame of rest" in the universe, it can only be the frame of reference of the atmosphere of neutrino particles or aether. It is however possible that our galaxy moves against its surrounding environment, where the cosmic microwave background comes from. This possibility will be further discussed later but whatever the case is, the anisotropy property of the microwave background has proven the presence of aether to transmit light!

Other findings have often been used by some physicists to prove that light travels at an absolute constant speed are from the astronomical observations of some binary star systems with their planes seen sideways from us, each star alternately moves toward and then away from us as it orbits around its companion star. These movements result in sending us alternative non-shift, red-shift, non-shift, and blue-shift spectra in exactly the stated sequence.

Astronomers have found such unscrambled spectra sequence from binary star systems. These observed results however only proved that all these lights travel at "the same speed" towards us, not even necessarily constant speed, and certainly not necessarily absolute constant speed.

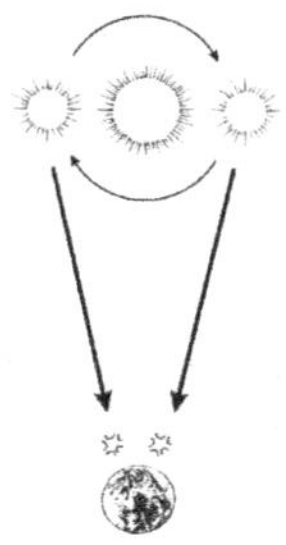

Lights coming from binary star systems travel at “the same speed” toward us but not necessarily at absolute constant speed or even at constant speed.

Some physicists religiously support their no-aether postulations to protect their modern physics and their 4-dimensional spacetime universe. Unfortunately, they are at the wrong side of science.

Scientific Origin of Light

Light is one of the most important universal phenomena like gravity. It should also be produced by the only atmospheric neutrino particles filling all the space of the universe. This is evident from the findings that neutrino particles travel at light speed and have small masses. From the close similarity between the physical properties of light to those of sound, it is very reasonable to draw the following conclusions:

a. Light is the particle wave of the atmospheric neutrino particles, which is the transmission medium of light.

b. Light also has a mechanical energy source. Unlike that air can transmit the vibration energies of a lot of objects to produce sound; neutrino particles only transmit the oscillation energies of electrons as light, likely nothing else.

c. Light travels at a constant, not an absolute constant, speed in space but only in a region where both the energy level and the concentration level of neutrinos are kept constant.

Doppler Effects

The Doppler Effect of sound was first investigated and explained by an Austrian physicist C. J. Doppler in 1842. He studied auditory phenomena, for example, the sound of a police car's siren grew higher in pitch as the car approached but lower as it drove away. He found that if there was relative motion between the sound source and the receiver, the frequency of the sound was altered: the frequency of wave-encounters were raised if the source and the observer were moving toward each other and lowered if they were moving apart. Again, it is an effect transmitted by a particle medium detectable by measuring the changes of wavelength, frequency and/or the speed of the mechanical wave caused by relative motions. A Doppler Effect provides quantitative information on relative moving velocities.

However, a change of the energy and/or concentration level of the transmission medium also causes change in wavelength, frequency, and the speed of the particle wave it transmitting without the presence of any relative motion. This is the so-called the transmission-medium Doppler Effects. Taking sound for example, both the temperature and the concentration of air are often lower on the top of the mountain than those at its bottom. When a sound travels from the bottom of the mountain to the top, its tune gets lower thus getting a red-shift Doppler Effect in the absence of relative motion. Besides, the sound gets weaker too, making observer believing that the source of the sound is farther away than it really is.

Doppler Effects are demonstrated and summarized using the figure and the table given below. When a light source S is moving toward stationary observer O and away from O', O detects a higher frequency waves than when S stays motionless and than what O' finds. These Doppler Effects are caused by the presence of relative motions between the light sources and the observers (case #1). Also, when S is its own observer, it can detect and measure its own motion against the transmission medium, by measuring its own dipole anisotropy (case #2). Besides, Doppler Effects can also occur when the energy and the concentration levels of the transmission medium have been changed (case # 3 and #4) in the absence of relative motion.

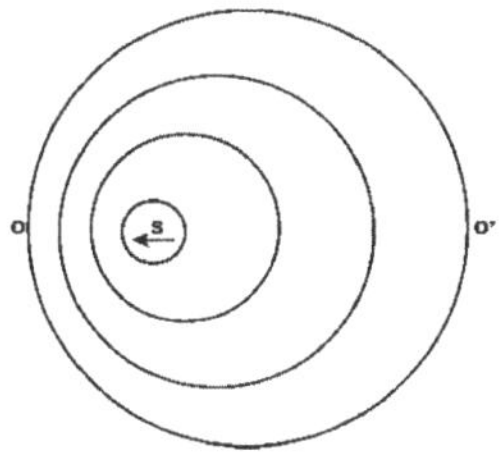

Doppler effects allow stationary observers O and O' to detect their relative motions. Besides, a light source S can also detect its own motion against the atmosphere of air, its transmission medium.

Case #	Light source (S)	Observer (O) or both (SO)	Medium energy & conc.	cause	Resulting Doppler Effect
1	S ↔ S S ↔	O O ↔ O ↔	-	Relative motion	Red or blue
2		SO↔	-	Motion in medium	Dipole anisotropy
3	S	O	↑ or ↓	Gravity	Red
4	S	O	↓ or ↑	Uneven star distribution	Red

Physicists insist that light does not have a transmission medium but still can have all the particle wave properties like sound including Doppler Effects. Not only that they do not have scientific evidence to support their claims but also that large amounts of experimental findings are against them! They believe that they can see light and observe Doppler Effects cases #1 and #3 only, not cases #2 and #4 due to that light has no transmission medium (Actually, without it, they can not even see light). According to them, the discovery of Doppler Effect cases #2 and #4 of light would prove the presence of a transmission medium in space to transmit light, thus, disprove modern physics. We however have already observed both Doppler Effects cases #2 and #4 of light! The dipole anisotropy property of the cosmic microwave background found by the COBE Satellite study of NASA in 1992 is for sure the case #2 Doppler Effect. The universal red-shifts in the remote star and galaxy lights are the case #4 Doppler Effect. Misinterpretation of these red-shift Doppler Effects has led to totally wrong modern cosmology! Why physicists believe that they can observe case #3 Doppler Effect? The answer is that #3 Doppler Effect was predicted by Einstein as gravitational red-shift from wrong scientific reasons, the presence of spacetime and curved spacetime instead of aether. The real scientific reasons are due to the presence of atmospheric neutrinos and its energy variations caused by their interactions with objects.

Universal Red-Shift and Big Bang Theory

Based on no-aether postulations physicists and cosmologists religiously believe that all the Doppler Effects of light are produced by relative motions having nothing to do with transmission medium. As a result, the scientific mistake in physics has led to a disastrously wrong cosmology.

In 1929 Hubble discovered red-shifts in the lights of all remote stars and galaxies and that the size of the red-shifts is proportional to their distances from Earth. This relationship is the so-called Hubble Law.

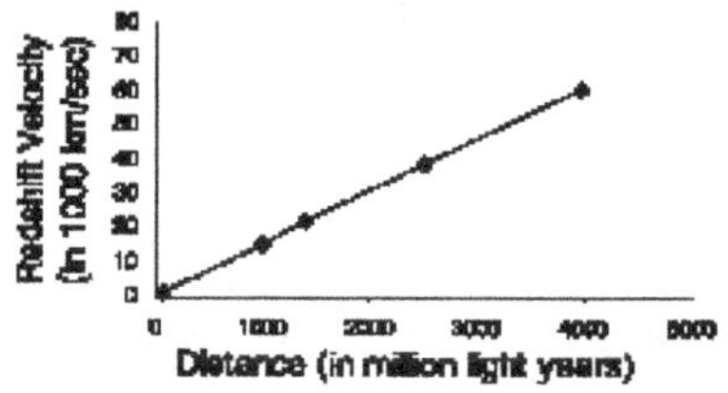

Red-Shift Velocity vs. Distance

Hubble Law plot: The size of the red-shift of a star or galaxy light is proportional to its distance from Earth.

The meaning and the importance of Hubble's findings to modern physics is summarized by David Wands: "The observational data available to Hubble by 1929 was sketchy, but whether guided by inspired instinct or outrageous good fortune, he correctly divined a straight line fit between the data points showing the red-shift was proportional to the distance. Since then much improved data has shown the conclusion to be a sound one. Galaxies are receding from us, and one another, as the Universe expands. Within General Relativity, the theory of gravity proposed by Albert Einstein in 1915, the inescapable conclusion was that all the galaxies, and the whole Universe, had originated in a Big Bang, thousands of millions of years in the past. And so the modern science of cosmology was born." The "inescapable conclusion" is of course based on the postulation that there is no aether in space to transmit light and unfortunately it was the wrong scientific conclusion.

Scientists have concluded from Hubble's findings and Hubble Law that the universe is uniformly expanding, thus growing. If we go back in time, the universe would get smaller, denser, and hotter. Eventually, if it does not become zero or nothing, it may become very small. At that time, it should be infinitely hot, dense, and have infinitely strong gravity. Big bang theory calls it "singularity" and postulated that the universe started with a big explosion of the singularity followed by expansion. The slope of the Hubble Law plot, thus, offers an estimation how old the universe is, about 15 billion years since the big bang.

Scientific Interpretation of Universal Red-Shift

1. Transmission Medium Effects

Both Hubble's findings of universal red-shift Doppler Effect in remote star and galaxy lights and Hubble Law have been confirmed by very large amounts of astronomical observations to be true. This Doppler Effect has two possible scientific causes: either due to the presence of relative motions or having transmission effects of light. With their no-aether postulations Einstein's theory of relativity and its modern physics have eliminated the possibility of transmission effects entirely! Now, Neutrinos have been found to be the aether responsible for producing all universal phenomena including transmitting light, producing universal forces, charged particles, etc. Besides being the only things filling the space of the universe, only neutrinos have the unique properties to explain and to predict the presence of all the universal phenomena! Without exception, both universal red-shift and Hubble Law are explainable by the presence atmospheric neutrinos. The following scientific interpretation has the support of very large amount of astronomical findings!

The most reliable astronomical findings are those obtained by telescopes. Using various telescopes astronomers have found billions and billions of stars about evenly distributed in the universe. At smaller scales however stars are not evenly distributed since they are inside galaxies only. Galaxies are about evenly distributed in the universe separated by vast starless intergalactic spaces (ISes). Since stars produce neutrinos, both the (kinetic) energy and the concentration of the atmospheric neutrinos should also be not evenly distributed like stars. They should be at maximum levels inside galaxies and at minimum levels at the center region of ISes. When a star light goes through an IS region, it slows down, gets both a red-shift and dimming effect. As shown in the following picture, light does not even travel at a constant velocity in space!

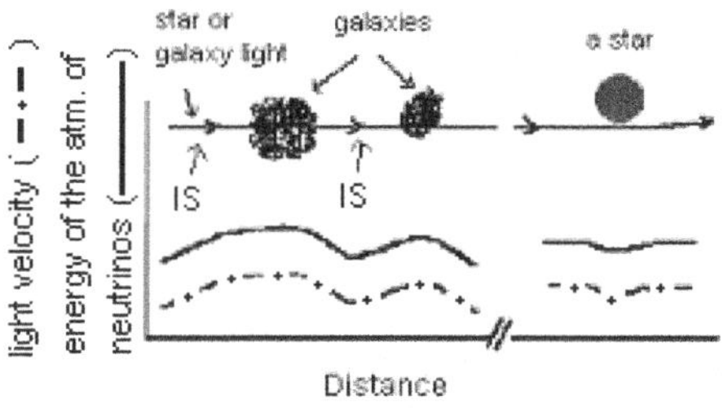

Light slows down and gets a red-shift upon passing through an IS region.

When a remote star or galaxy light travels towards Earth, it goes through many galaxies and ISes. Since neutrinos have higher energy and concentration inside galaxies than inside ISes, a star light goes faster inside galaxies and slower inside ISes. A star light gets both a red-shift and dimming effect every time when it goes through an IS and both effects are additive. Due to the variation of energy and concentration levels of the atmospheric neutrinos, a star light does not even travel at a constant velocity in space as shown in the above picture! Similarly, a star light also slows down and gets a red-shift upon passing by a massive star or planet.

In conclusion, a Doppler Effect has two possible causes: by relative motions or by transmission medium effects. Scientists must rely on observatory findings to determine which one is the real cause. The presently accepted modern physics and modern cosmology are however based on no-aether postulations and, therefore, has unscientifically removed the scientific possibility of transmission medium effects leaving the only possible cause for the universal red-shift to be relative motions. Now aether has been found to be neutrinos and the cause of the universal red-shift has been determined to be the transmission medium effects of the atmospheric neutrinos. The unscientific no-aether postulations have led to scientifically wrong modern physics and modern cosmology!

2. New Meaning of Hubble Law and its Applications

When a star or galaxy light passes through an IS, besides getting a red-shift, it simultaneously gets a dimming effect from the lower energy and diluter transmission atmospheric neutrinos.

Astronomers have long been using the method of light brightness comparison of certain types of stars to measure the distances of remote stars and galaxies. However, the results so obtained should have a systematical error of over estimation due to that they have not taken the light-dimming effect of the ISes into consideration. Fortunately, the light-dimming effect should be proportional to the red-shift effect and both effects should be the same measurement of the total energy reduction of the (star and galaxy) lights from their passing through all ISes. Since Hubble Law shows that the red-shift (total energy reduction) of star and galaxy lights is also proportional to their distances from us, combining all the above three proportional relationships we have a very accurate method to measure the distances of remote stars and galaxies. The total energy reduction (TER) of a red-shifted light can be calculated by $[(\upsilon - \upsilon')/ \upsilon] \times 100\%$. Where υ is a frequency of the none-shifted light; υ' is the corresponding frequency of a red-shifted light. The correction of the systematic error of overestimation is therefore "TER x measured distance" by a star light brightness comparison method. The universal red-shift data therefore can be used to correct the systematic error of the measured distances of stars and galaxies obtained by all the light brightness comparison methods.

Presently, the distances of remote stars, galaxies, and the visible size of the universe have all been overly estimated and should be corrected!

Energy Source and Backgrounds of Light

There are very large amounts of experimental findings to conclude that the oscillation of electrons, largely in atomic and molecular orbital, produces light. As pointed out earlier that both proton and electron have their own atmospheres of neutrinos to greatly enhance their efficiency of energy-exchanging interactions with the atmospheric neutrinos of the universe. Although all charged particles may be capable of producing light, electron is unique in having the densest atmosphere of neutrinos among all charged particles commonly present such as electrons, protons and nuclei. Having a nature environment to oscillate at high frequencies is another difficult condition for other charged particles to produce light. Although it is theoretically possible for all other charged particles such as proton and nuclei of atoms to also produce lights, scientists have not found them yet.

Therefore, it is very unique that only the oscillating electrons in atoms, molecules, nuclei, etc. are the main energy sources of light. Yet, light has a very broad energy range of spectrum and the visible range is only a very narrow portion of it. In the visible range alone, orbital electrons have thousands of finely separable energy levels that make a very colorful universe. For us to see everything in such detail even under microscope there must be a very dense atmosphere of neutrino particles. Since only matters made of atoms and molecules have both orbital and nuclear electrons in nature states readily to produce and to reflect lights at their own special energy levels. We therefore can only see matters having atomic and molecular structures.

Without having atomic structure dense-matter objects likely do not radiate nor reflect light. However, since they have compositions and structures similar to those of nuclei of atoms, they should radiate very high frequency lights. For example, we have detected x-rays from suspected black holes and neutron stars, etc.

Neutrino Winds and CMB

We have detected cosmic microwave background (CMB) having characteristics of black body radiation indicating that it is radiated by matters having atomic and molecular structure at very low temperature (~2.7 ^{o}K). Having billions of stars distributed over all our galaxy space, the planets and asteroids inside our galaxy unlikely have such a uniformly low temperature. Only the matters inside an intergalactic space (IS) region surrounding our galaxy may have such a uniformly low temperature and is likely the source of CMB. All ISes are likely the dump ground for galaxies and may have high contents of planets, dead stars, and asteroids.

In 1992 NASA found anisotropy property in their CMB mapping study. Today's physicists support big bang theory believing CMB is the residue radiation predicted by the big bang. Some physicists however suggested finding anisotropy property in CMB might mean discovery of aether and that Earth or the entire galaxy are moving in it. According to the anisotropy data, Earth is moving in aether at an absolute velocity around 600 kilometers per second.

Finding aether has already disproved all no-aether theories including big bang theory and therefore its explanation of CMB is also wrong. The above suggested scientific explanation that Earth may encounter a strong neutrino wind should be provable and disprovable experimentally. In the presence of such a strong wind of neutrinos, all the light sources on Earth should also have an anisotropy property having the mirror image of the CMB background.

The picture given below shows an exaggerated anisotropy property of all the light sources on Earth predicted to have. Darker color indicates shorter wavelength and lighter color represents longer wavelength.

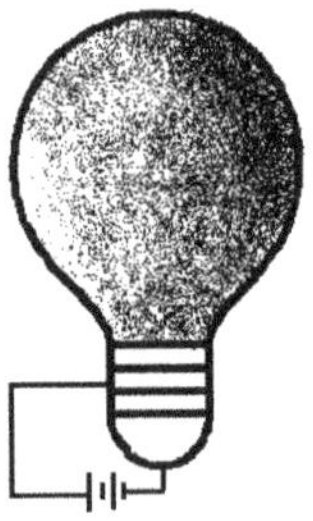

Moving against a strong wind of neutrinos, all the light sources on Earth should have an anisotropy property having the mirror image of CMB.

However, no such anisotropy property on any light source has been reported. Besides, in the presence of such a strong neutrino wind, the rotational and the orbiting motions of the Earth, the gravitational force on Earth should be uneven and change periodically daily. Apparently, this is also not true. There are therefore strong scientific evidences against the suggestion of the presence of a strong neutrino wind against Earth.

A new scientific interpretation on the anisotropy property of the CMB is needed and is given here. Instead of moving against "aether", more likely, our Earth or the entire galaxy may move with its neutrino wind. The anisotropy found in CMB shows that our galaxy as a whole is moving against something (the so-called frame of reference) radiating CMB surrounding our galaxy. Since

CMB is produced by atomic and molecular matters having very uniformly low temperature, ~2.7°K, the CMB producing matters can not exist inside our galaxy or any galaxy where billions of stars live. There is however a vest starless intergalactic space (IS) surrounding our galaxy and it likely has many dead planets and stars. Without stars, all matters inside IS likely have unusually uniform and low temperature! It is therefore possible that our galaxy as a whole is moving in a spiral neutrino wind, which also moves against its surrounding IS, which constantly radiates CMB.

The remaining question is why CMB is smooth instead of granular like the telescopic findings of stars? The following factors may have contributed to the smoothness of CMB. 1. The extremely low temperature in the intergalactic space (~2.7 °K) may have shattered all crystal structures of matters causing dead planets and stars to break up. 2. It is likely that the IS is a very windy place since there are many neutrino winds constantly blowing from different galaxies into it. Strong neutrino winds therefore may also contribute to the breaking up of dead planets and stars there. 3. Experimental method may have something to do with the smoothness of the CMB. CMB is not observed by telescopes. Measuring weak energy may need an area instead of a spot. The method likely takes an average value from an area, thus, contributes to the smoothness of CMB.

Breaking up dead stars and planets into pieces would greatly enhance the CMB radiation in ISes and the smoothness of their CMB signal too.

Chapter 3 – Cosmology

So far PM essentially has nothing to offer on the scientific topics of universal phenomena and the universe beyond its old prediction that there must be matters (aether) taking up all space to produce light and universal forces. The presently accepted universal views are those of modern physics based on several no-aether theories. Modern cosmology is also based on Einstein's theory of general relativity and big bang theory and both are no-aether theories. Both of them were wrong to begin with since their fundamental postulations have violated the science of PM. Upon the discovery of aether, now the PM-based new aether science already offers a broader range scientific interpretation of all universal phenomena even han the combined scope of all no-aether theories. Unfortunately, the new aether science is fundamentally incompatible with all no-aether theories of course including modern physics and modern cosmology, which have long been accepted and supported by the scientific community and taught in schools and colleges.

Modern cosmology is based on many strong beliefs on no-aether postulations. For example, we were taught to believe that the universe started from a big bang and stars are formed from gravitational pull among hydrogen and helium molecules. We also believe that there is no aether in the universe to produce light and universal forces. We believe that light speed is absolute but time and dimension are not. We even believe that all universal phenomena are mathematical equations and even the universe should be interpreted (ruled) by mathematical equations. In the mathematical model of modern cosmology gravitational force is the only major force to shape the universe. It is the force to form stars, to start and maintain nuclear reactions at the center of stars. Gravity can be infinitely strong so that it can crash molecular and atomic structures, thus, making dead stars having amazingly dense masses. Although the above beliefs have long been accepted by the scientific community and are taught in schools and colleges, they are part of the theory based on modern physics or Einstein's theory of relativity not scientifically proven correct. Even if they were true, there many unanswered scientific questions. For example, what kind of nature engine inside the center of all stars that can continue to "pump" unused hydrogen to the center and to "purify" and to "remove nuclear-reaction wastes" to maintain constant nuclear reaction rates for billions of years!

Finding aether has completely disproved modern cosmology. Fortunately, the new aether science is ready to replace it.

A. Fundamental Universal Views

Theory of general relativity including big Bang theory is the scientific foundations of modern cosmology. The discovery of aether has disproved it and brought an entirely new physics of the universal phenomena and scientific cosmology. New aether science offers a mechanical, not mathematical, explanation of the universal phenomena, which is consistent with PM. The mechanical universe is still the old 3-dimensional universe before Einstein, where both time and dimensions are absolute but not light speed.

Although modern cosmology has been developed based on some astronomical observations and well established knowledge of physics, chemistry, etc., it is a mathematical model similar to that of Ptolemy unable to offer scientific interpretation to any of the universal phenomena. The new aether science offers a scientific interpretation for all universal phenomena. Upon scientifically understanding them, it is easy for scientists to determine what mathematics to use. After centuries of focusing on mathematics, certainly some of them are correct and should be utilized by new aether science.

The Domain of the Nature (DON)

A universe can be defined as a place where large amounts of stars live. Stars constantly release huge amount of energetic neutrinos, light and heat energies. The universe is therefore an energy consuming system like an engine of a running car or a fire constantly burning fuel. There is always limited amount of fuel, thus, limited duration for any energy consuming system. Besides, the energy of an energy-consuming system must be produced and stored first by some nature ways before it can be consumed. Also, there should be a nature mechanism to trigger the starting to consume the stored energy. From what we know, the nature has designed a perfect engine inside all stars to continuously work for billions of years maintenance-free to burn naturally stored energy by nuclear reactions at essentially constant rates!

For a fire to burn there must be a bigger environment to make and store its fuel. For example, Earth provides an environment for plants to receive radiation energy from the Sun to be born, to grow old, and to die. A lightening would trigger the starting of a tiny forest fire, which spreads and, eventually, when the stored forest fuel, the plants dead or alive, is exhausted, the fire dies.

A universe must also be living in an even bigger and longer-lasting environment to store energy and to be born (to start consume stored energy), to grow, and to die in it. The author calls this bigger environment the domain of the nature (DON), which is the bigger environment to collect and store energy producing fuel for a universe to be born, live, and die. To understand our universe, we need to figure out some basic properties about the domain of the nature too.

The knowledge about our universe should be a good start to figure out what the DON should be, which appears to have unlimited space and time, for example, but not the universe. There may be many universes inside the DON obeying the same laws of physics and chemistry, existing simultaneously and/or at different time periods. The DON may be similar to our universe having many galaxies. Of course a universe is separated by much greater starless space than an intergalactic space (IS) from other universes. The atmospheric neutrino particles of all universes continuously escape into the DON or the spaces separating the universes. Leaving a universe into the vast starless space of the DON, neutrino particles continue to lose their kinetic energy and to reduce their concentrations to very low levels. There is however a dilute weak energy atmosphere of neutrinos surrounding a universe and therefore there are forces of DON, which are similar to the universal forces but are much weaker. The electrons and protons in DON may still be charged particles but their charges are much weaker than those in a universe. These weakly charged particles may not be able to form atoms and molecules having nuclei and orbital electrons. Instead, under the weak forces of DON they may form weakly bonded aggregates, which continue to grow in sizes by collecting surrounding electrons and protons.

In summary, many universes may co-exist inside the domain of the nature and there is likely a very dilute and low energy atmosphere of neutrinos in the DON surrounding all universes. The energy and the concentration of the atmospheric neutrinos are likely so low there that no matter having atomic and molecular structures exist and therefore no light can exist and be transmitted through there. It means that we are unable to detect the existence of other universes by light or any other technologies based on matters made of atoms and molecules. There are however "forces of the DON" produced by the very weak and dilute atmosphere of neutrinos, which are much weaker than the universal forces. These forces aggregate electrons and protons to form dense-matter objects, which giving time can grow very big.

In the DON the atmosphere of neutrinos should be unevenly distributed and there may be places where real vacuum exist without neutrino particles in gas-like form.

Open Universe

Contrary to general belief, like a fire, the mass and the energy of a universe is not conserved. The stars of a universe constantly produce large amount of neutrinos, which continuously escape out of it carrying mass and energy. It is a big scientific mistake for theory of relativity and modern physics to postulate that the universe is an enclosed system having conserved mass and energy.

Dense-Matter Objects

Upon realizing that atomic structure can not exist outside universes, a universe can not begin with matters having atomic structures! It means that at the very beginning our universe is made of some kind of matter unknown to us!

Both the energy and the concentration of the atmospheric neutrinos continue to decrease upon leaving a universe. The temperature of the IS surrounding our galaxy is kept at ~2.7 ^{o}K by all its surrounding galaxies. However, the temperature outside a universe will continue to drop to 0 ^{o}K and below. Scientists have found that upon cooling to very close to 0 ^{o}K, orbital electrons of atoms collapse into their nuclei to form a very dense matter yet unknown to us. The author has taken the liberty to name it the "**dense-matter object**". The findings proved that outside universe atoms and molecules can no longer exist by collapsing into dense-matter objects. Without shinning star, the temperature of a dying universe would continue to drop to 0 ^{o}K and below. Then, its matters having atomic and molecular structures will collapse to form dense-matter objects. Still under a low-energy dilute atmosphere of neutrinos, dense-matter objects would tend to agglomerate and continue to grow in size. Giving time, there is essentially unlimited supply of time in the DON, some of the dense-matter objects can grow to very large size having amazingly large masses and relative kinetic energies.

Under very weak dilute atmosphere of neutrinos, protons and electrons may still be weakly charged and still have their own atmospheres of neutrinos. Low energy neutrinos may be even easier to be caught by the atmospheres electrons and protons and they may be heavier and having denser atmospheres of neutrinos than those inside a universe under higher energy atmosphere of neutrinos. Dense-matter objects therefore may be made of protons and electrons having enriched neutrino atmospheres of their own, thus, may be even denser than atomic nuclei. There are therefore dense-matters objects formed and stable outside the universe and in dead universes. Like atoms, they are made of protons and electrons but their proton-electron composition ratio may vary somewhat around the fixed ratio of 1:1 like those in atoms.

Having extremely dense mass, even under very weak dilute atmosphere of neutrinos, a large size dense-matter object may still have quite strong gravitational force. It therefore continues to grow in size by collecting surrounding subatomic particles and dense-matter objects. Giving time, some of them can grow to amazingly large sizes. Making large sizes of dense-matter objects appears to be the nature way to concentrate both matter and energy needed to make new galaxies and universes.

Having similar structure to atomic nuclei without orbital electrons to protect them, we would expect that dense-matter objects are unstable inside a universe like radioactive materials. Therefore, inside a universe under constant bombardment of energetic neutrinos, dense-matter objects should be unstable becoming "radioactive". However, since they have amazingly dense mass, the atmospheric neutrinos can not penetrate deep into them. Therefore, inside universe only their surfaces are constantly bombarded by atmospheric neutrinos causing constant releasing of electrons, protons, neutrinos along with huge amount of energy, which initiate and maintain constant rates of nuclear reactions only on their surface. The DON therefore has the best nature way to concentrate mass, store nuclear energy, and having the best sustained-release method to utilize it inside a universe!

What the DON is doing is to concentrate masses by making ever growing size dense-matter objects, which are rich in (relative) kinetic energy and nuclear energy ready to be release inside a universe. All universes therefore must have an energetic atmosphere of neutrinos to utilize the dense-matter objects to make atomic matters and to utilize the huge energy they have stored. The dense-matter objects therefore should be the starting materials for making the stars of universes like ours. Since matters having atomic and molecular structures can not exist in DON outside universes, a universe can not be born with them. Therefore, a universe should be born to contain 100% dense-matter objects.

Formation of a Galaxy and a Universe

A large dense-matter object has amazingly large and dense mass. It therefore has amazingly large kinetic energy relative to other dense-matter objects. If dense-matter objects are the starting material of the universes, the remaining question is how a galaxy or universe gets started? Findings suggest that most stars in a galaxy appear to have about the same age indicating that billions of stars inside a galaxy may be born simultaneously. It means that at the beginning, there were already billions of pieces of dense-matter objects to begin nuclear reactions simultaneously on their surfaces to make hydrogen and helium to become stars.

The nature way to start a galaxy therefore is likely a violent collision between two huge dense-matter objects as illustrated in the picture given below:

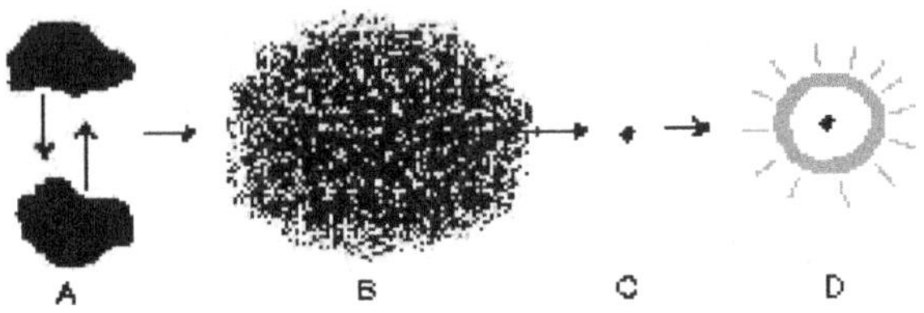

A violent collision of two huge dense-matter objects (A) would break them into billions of pieces and produce a dense cloud containing dust of dense-matter object, electron, proton, neutrinos and huge amount of energy (B). Under constant bombardment of the energetic cloud particles, the broken pieces of dense-matter objects and their dust particles become "radioactive" starting nuclear reactions on their surfaces. The nuclear reactions have quickly consumed the dust and the small size dense-matter object pieces producing hydrogen and helium clouds and an energetic atmosphere of neutrinos. Meanwhile, all the large broken pieces of the dense-matter-objects quickly become baby stars by having continuous nuclear reactions on their surfaces to produce protons, electrons, hydrogen, helium, and neutrinos. The simultaneous formation of billions of baby stars marks the birth of a baby galaxy (B). These newly born baby stars although very hot are invisible since they contains essentially 100% dense-matter object, which without having atomic and molecular structures do not produce and reflect light (C). An invisible baby star contains ~100% dense-matter object and is an antigravity star since it shots out energetic subatomic particles and neutrinos. Therefore, at the beginning a baby galaxy uniformly expands due to that all its baby stars are antigravity stars shooting subatomic particles and neutrinos at one another. A baby star also shoots out the hydrogen and the helium it continuously makes from its nuclear reactions to a great distance out due to its antigravity property. Giving time the hydrogen and the helium particles gradually accumulate around a baby star and with the help of their own gravity and other universal forces form an outer layer, which is at a great distance from the dense-matter object at its center and continuously gets thicker. As the outer layer grows thicker, the baby star gets brighter (D).

Having gravitational force affecting one another, many large dense-matter objects may exist in group environments having orbiting-like motions. When two groups of dense-matter objects run into each other, chances are many collisions will happen over a long period of time with each collision forming a galaxy. A universe may form this way over a rather long period of time. Galaxies in a universe therefore have different ages. Some are young and some old. Some may have been dead before a new galaxy is born. A universe therefore does not have a meaningful age but its individual galaxies do. A galaxy will die when its dense-matter objects are about to be used up. A group of dense-matter objects may run into a universe and sometimes a large dense-matter object may collide with stars to make more new stars giving out unbelievably high-energy and explosion-like views.

B. Stars

Astronomical Findings and Modern Cosmology

Under no-aether postulations the science and the mathematical models of modern cosmology for all stars are wrong. Scientifically a singularity having infinitely strong gravity, high temperature, and mass density can not be naturally formed and therefore can not exist. Also, stars can neither be naturally formed from gravitational pull of dispersed hydrogen and helium gases nor gravitational force can start nuclear reactions at the center of stars where no gravity is present.

Modern cosmology is however also based on large amount of astronomical observations, the knowledge of physics and chemistry. It is therefore useful to review modern cosmology and astronomy before introducing the new cosmology of new aether science, which is still in its infancy stage having essentially only one man's view. A brief introduction and comparison between the two is given below. The modern cosmology given here can be found in basic cosmology text books and from Internet scientific educational websites and is therefore considered proven, well established, and accepted by the scientific community.

1. Stars Formation

According to modern cosmology gravitational force can be infinitely strong and is the force to collect hydrogen and helium clouds to form stars. Strong gravitational force would generate such high temperatures at the center of stars to start and to maintain their nuclear reactions there. Also, strong gravity would crash atoms and molecules to form very dense fused-masses such as the formation of neutron stars and black holes.

Modern cosmology offers a no-aether postulations-based interpretation of the universe, which is based on modern physics' interpretation of the universal red-shift of star and galaxy lights and Hubble Law. According to modern cosmology the universe began about 15 billion years ago from a "big bang", a cosmic explosion of a singularity that resulted in an expanding cloud of the two lightest elements, the gases of hydrogen and helium. Where there were higher concentrations of gases, the mutual gravitational attractions of the gas molecules led to the growth of the first generation of stars! As more and more material fell into a new star, the pressure at its center finally became high enough to start the process of nuclear fusion, in which the nuclei of hydrogen and helium merge to form heavier elements. This was accompanied by the release of energy, which made the star beginning to shine. Eventually, all the hydrogen and helium, and those products that could be used to generate energy in the core of the star, were exhausted, and the nuclear furnace was extinguished. As a result, the outer layers of the star could no longer resist the central force of gravity, which was pulling the star's outer matter inward toward its core. What happened next depended on the mass of the star. Very large stars become supernovas, exploding violently spewing much of the stellar material into space and their very strong gravity then crash the atoms and molecules of these stars into very dense fused mass known as black holes and neutron stars. For small stars the process was slower; instead of an explosion, elements from the star's interior zones rose to the surface and were then lost to space when the outer layers blew off, again, leaving strong gravity compressed dead stars known to be white and brown dwarfs.

Postulating no aether in space to transmit light and to produce universal forces is the fatal mistake of both modern physics and modern cosmology. For example, the wrong interpretation of Hubble's findings of universal red-shift in star and galaxy lights and Hubble Law has led to the development of big bang theory, on which the modern cosmology is based. It is very hard to believe that dispersed hot hydrogen and helium gases can be collected by their own gravitational force to form stars. Even if stars are formed from gravitational pull and stars have enough hydrogen fuel to undergo nuclear reactions for billions of years, cosmologists still can not explain how stars can continue to extract pure hydrogen and bring it to their centers and to remove nuclear wastes from its center to maintain essentially constant nuclear reaction rates for billions of years.

Assuming that all matters have attraction force capable of acting from distance is very unscientific. PM neither supports attraction force nor recognizes acting-from distance force. If all matters have gravitational force, they should have a source constantly supply them with energy to create and maintain gravity. So far, no such energy source has been identified. A (gravitational) pressure does not automatically and constantly produce heat. Based on our knowledge, it may cause chemical and physical changes, which either release or absorb heat. These changes only occur over a short period of time in comparison to astronomical time and the heat energies they produced are too low to ignite nuclear reactions. If gravitational pressure provides heat to ignite and maintain nuclear reactions, the nuclear reactions in stars should occur at a quarter distances inside stars, where their gravitational forces have reached maximum, not at their centers, where no gravitational force exists!

Based on astronomical findings modern cosmology suggested that large and medium size stars explode to end their lives upon using up their hydrogen and helium fuels and nuclear reactions keeps the strong gravitational pull of stars to collapse them. If it is true, stars should collapse to end their lives instead of exploding up.

2. Neutron Stars

Neutron stars are about 20 km in diameter and have the mass of about 1.4 times that of our Sun. They are so dense that one teaspoonful would weigh a billion tons on Earth! Because of its small size and high density, a neutron star possesses a surface gravitational field about 2 x 10^{11} times that of Earth. Neutron stars also have magnetic fields a million times stronger than the strongest magnetic fields produced on Earth.

Neutron stars are one of the possible ends for a star. They result from massive stars which have mass greater than 4 to 8 times that of our sun. After these stars have finished burning their nuclear fuel, they undergo a supernova explosion. This explosion blows off the outer layers of a star into a beautiful supernova remnant. The central region of the star collapses under gravity. It collapses so much that protons and electrons combine to form neutrons, hence the name "neutron star".

Neutron stars may appear in supernova remnants, as isolated objects, or in binary systems. One neutron star is thought to have planets. When a neutron star is in a binary system, astronomers are able to measure its mass. From a number of such binaries seen with radio or X-ray telescopes, the neutron stars were found to have masses of about 1.4 times that of the Sun. For binary systems containing an unknown object, this information helps distinguish whether the object is a neutron star or a black hole, since black holes are more massive than neutron stars.

Pulsar - Pulsars are rotating neutron stars. And pulsars pulse because they rotate! Pulsars were first discovered in late 1967 by graduate student Jocelyn Bell Burnell as radio sources that blink on and off at a constant frequency. Now we observe the brightest ones at almost every wavelength of light. Pulsars are spinning neutron stars that have jets of particles moving almost at the speed of light streaming out above their magnetic poles. These jets produce very powerful beams of light. For a similar reason that "true north" and "magnetic north" are different on Earth, the magnetic and rotational axes of a pulsar are also misaligned. Therefore, the beams of light from the jets sweep around as the pulsar rotates, just as the spotlight in a lighthouse does. Like a ship in the ocean that sees only regular flashes of light, we see pulsars turn on and off as the beam sweeps over the Earth. Neutron stars for which we see such pulses are called "pulsars", or sometimes "spin-powered pulsars," indicating that the source of energy is the rotation of the neutron star. Some pulsars also emit X-rays.

3. White Dwarf Stars

White dwarf is produced when a low to medium mass star dies. These stars are not heavy enough to generate the core temperatures required to fuse carbon, and after they have become a red giant during their helium-burning phase, they will shed their outer layers to form a planetary nebula, leaving behind an inert core consisting mostly of carbon and oxygen.

This core has no further source of energy, and so will gradually radiate away its energy and cool down. The core, no longer supported against gravitational collapse by fusion reactions, becomes extremely dense, with a typical mass of about half that of the sun contained in a volume about equal to that of the Earth. Many white dwarfs are approximately the size of the Earth, typically 100 times smaller than the Sun. They may have the same mass as the Sun and so are very compact. A radius which is 100 times smaller, implies that the same amount of matter is packed in a volume that is typically 100^3=1,000,000 smaller than the Sun and so the average density of matter in white dwarfs is 1,000,000 times denser than the average density of the Sun.

Eventually, over hundreds of billions of years, white dwarfs cool to temperatures at which they are no longer visible. However, over the universe's lifetime to the present (about 15 billion years) even the oldest white dwarfs still radiate temperatures of a few thousand Kelvin's.

White dwarf stars are extremely hot; hence the bright white light they emit. This heat is a remnant of that generated from the star's collapse, and is not being replenished (unless it accretes matter from other nearby stars). However, since white dwarfs have an extremely small surface area from which to radiate this heat energy, they remain hot for a long period of time. Eventually, a white dwarf will cool into a black dwarf. Black dwarfs are ambient temperature entities and radiate weakly in the radio spectrum, according to theory. However, the universe has not existed long enough for any white dwarfs to have cooled down this far yet, so no black dwarfs are thought to exist. Many nearby, young white dwarfs have been detected as sources of soft X-rays (i.e. lower-energy X-rays); soft X-ray and extreme ultraviolet observations enable astronomers to study the composition and structure of the thin atmospheres of these stars.

White dwarfs cannot be over 1.4 solar masses, the Chandrasekhar limit, but there is a working method to get them over this limit. A white dwarf can accrete material from a companion. Unlike a nova, the material accretes slowly and remains stable. The mass of the white dwarf increases until it hits the 1.4 solar mass limit, at which degeneracy pressure cannot support the star. This creates a type Ia supernova and is the most powerful of all the supernovae.

White dwarf was discovered in 1862 by Alvan Graham Clark by discovering a dark companion of the brightest star Sirius (Alpha Canis Majoris). The companion, called Sirius B or the Pup, had a surface temperature of about 25,000 K, so it was classified as a hot star. However, Sirius B was about 10,000 times fainter than the primary, Sirius A. Since it was very bright per unit of surface area, the Pup had to be much smaller than Sirius A, with roughly the diameter of the Earth.

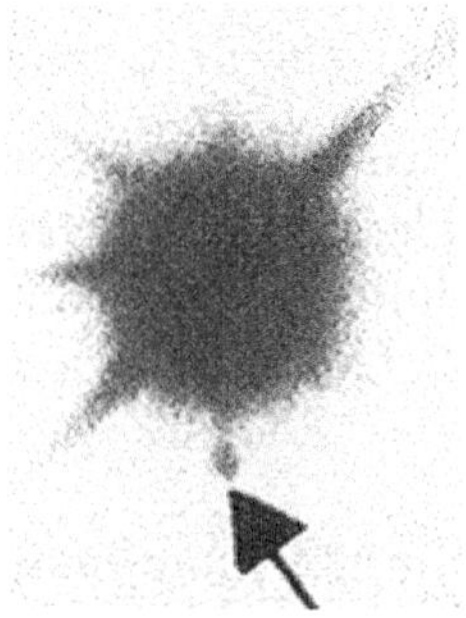

The arrow points to Sirius B, a white dwarf star.

Analysis of the orbit of the Sirius star system showed that the mass of the Pup was almost the same as that of our own Sun. This implied that Sirius B, a white dwarf, was thousands of times more dense than lead. Later, astronomers found that white dwarfs are common in our galaxy.

4. Brown Dwarf Stars

Brown dwarfs are substellar objects (~5 to 90 Jupiter masses) that do not fuse hydrogen into helium in their cores, as do stars on the main sequence, but have fully convective surfaces and interiors, with no chemical differentiation by depth. There is some question as to whether brown dwarfs are required to have experienced fusion at some point in their history; in any event, brown dwarfs heavier than 13 Jupiter masses do fuse deuterium.

Early stellar models suggested that a true star requires a mass at least 80 times that of Jupiter to support such fusion. Dense star-like objects with smaller masses, or "brown dwarfs," were hypothesized by the early 1960s -- formed much the way stars are formed, they would however be hard to find in the sky, as they would emit almost no light. Their strongest emissions would be in the infrared (IR) spectrum, and ground-based IR detectors were too imprecise for a few decades after that to firmly identify any brown dwarfs.

More recently, it has been hypothesized that, depending on the compounds that make up a growing stellar object, the critical mass for star-like hydrogen fusion could be as large as 90 Jupiter masses; and on the other end of the spectrum, that substellar objects formed quickly from a collapsing nebula could produce brown dwarfs smaller than 13 Jupiter masses, which nevertheless experience no fusion at all.

Since 1995, when the first brown dwarf was confirmed, hundreds have been identified. They are now believed to be the most numerous type of body in the Milky Way. Brown dwarfs close to Earth include Epsilon Indi Ba and Bb, a pair of dwarfs around 12 light-years from Sun.

Mainline stars cool, but eventually reach a minimum luminosity which they can sustain through steady fusion. This varies from star to star, but is generally at least 0.01% the luminosity of our Sun.

Brown dwarfs cool and darken steadily over their lifetimes: sufficiently old dwarfs will be too faint to be a star.

A remarkable property of brown dwarfs is that they are all roughly the same radius, more or less the radius of Jupiter. At the high end of their mass range (60-90 Jupiter masses), the volume of a brown dwarf is governed primarily by electron degeneracy pressure, as it is in white dwarfs; at the low end of the range (1-10 Jupiter masses), their volume is governed primarily by Coulomb pressure, as it is in planets. The net result is that the radii of brown dwarfs vary by only 10-15% over the range of possible masses. This can make distinguishing them from planets difficult.

In addition, many brown dwarfs undergo no fusion; those at the low end of the mass range (under 13 Jupiter masses) are never hot enough to fuse even deuterium, and even those at the high end of the mass range (over 60 Jupiter masses) cool quickly enough that they no longer undergo fusion after something on the order of 10 million years. However, there are other ways to distinguish dwarfs from planets: Density is a clear giveaway. Brown dwarfs are all about the same radius and volume; so anything that size with over 10 Jupiter masses is unlikely to be a planet. X-ray and infrared spectra are telltale signs. Some brown dwarfs emit X-rays; and all "warm" dwarfs continue to glow tellingly in the red and infrared spectra until they cool to planet like temperatures (under 1000K).

Typical atmospheres of known brown dwarfs range in temperature from 300 to over 3000 K, in comparison with stars, which cool to minimum temperatures of around 4000 K. Compared to stars, which warm themselves with steady internal fusion, brown dwarfs cool quickly over time; more massive dwarfs cool slower than less massive ones.

5. Black Holes

A black hole is a region of spacetime from which nothing can escape, even light. It is impossible to see a black hole directly because no light can escape from them; they are black. But there are good reasons to think they exist.

When a large star has burnt all its fuel it explodes into a supernova. The stuff that is left collapses down to an extremely dense object known as a neutron star. We know that these objects exist because several have been found using radio telescopes. If the neutron star is too large, the gravitational forces overwhelm the pressure gradients and collapse cannot be halted. The neutron star continues to shrink until it finally becomes a black hole. This mass limit is only a couple of solar masses, that is about twice the mass of our sun, and so we should expect at least a few neutron stars to have this mass. (Our sun is not particularly large; in fact it is quite small.)

A supernova occurs in our galaxy once every 300 years, and in neighboring galaxies about 500 neutron stars have been identified. Therefore, we are quite confident that there should also be some black holes.

Below is the Hubble Space Telescope image of the center of galaxy NGC4261. Inside it, there is a brown spiral-shaped disk that is rotating. Based on gravitational calculations, it weighs hundred thousand times as much as our Sun. Physicists believe that there is a black hole at its center having gravity about one million times that of our Sun.

Hubble Space Telescope image of the center of galaxy NGC4261. Physicists believe that there is a black hole at its center.

6. Dark Matters

Cosmologists believe that most of the matter in the universe is *dark*, i.e. cannot be detected from the light which it emits (or fails to emit). This is "stuff" which cannot be seen directly -- so what makes us think that it exists at all? Its presence is inferred indirectly from the motions of astronomical objects, specifically stellar, galactic, and galaxy cluster/supercluster observations. It is also required in order to enable gravity to amplify the small fluctuations in the Cosmic Microwave Background enough to form the large-scale structures that we see in the universe today.

For each of the stellar, galactic, and galaxy cluster/supercluster observations the basic principle is that if we measure velocities in some region, then there has to be enough mass there for gravity to stop all the objects flying apart. When such velocity measurements are done on large scales, it turns out that the amount of inferred mass is much more than can be explained by the luminous stuff. Hence we infer that there is dark matter in the Universe.

Dark matter has important consequences for the evolution of the Universe and the structure within it. According to general relativity, the Universe must conform to one of three possible types: open, flat, or closed. The total amount of mass and energy in the universe determines which of the three possibilities applies to the Universe. In the case of an open universe, the total mass and energy density (denoted by the Greek letter Omega) is less than unity. If the Universe is closed, Omega is greater than unity. For the case where Omeg*a* is exactly equal to one the Universe is "flat".

Note that the dynamics of the Universe are not determined entirely by the geometry (open, closed or flat) unless the Universe contains only matter. In our Universe, where most of Omega comes from dark energy, this relation between the mass density, spatial curvature and the future of the universe no longer holds. It is then no longer true in this case that "geometry (spatial curvature) is destiny." Instead, to find out what will happen, one needs to calculate the evolution of the expansion factor of the universe for the specific case of matter density, spatial curvature and "funny energy" to find out what will happen.

Dark matter (DM) candidates are usually split into two broad categories, with the second category being further sub-divided into Baryonic, Non-Baryonic, hot dark matter (HDM), and cold dark matter (CDM) depending on their respective masses and speeds. CDM candidates travel at slow speeds (hence "cold") or have little pressure, while HDM candidates move rapidly (hence "hot").

Current indications from the cosmic microwave background are that the universe is spatially flat. That implies that the sum of all of the energy (density) in the universe equals the critical density, i.e. the total Omega is 1. This is quite interesting because as the universe expands the value of Omega changes. In fact the value 1 is unstable, and the universe would prefer to evolve towards one of the two natural values: 0, if the universe expands forever further apart until the universe is almost totally empty; and infinity, if the matter re-collapses to a state of higher and higher density. Then the observation that Omega is fairly close to 1 today, means that it must have been even closer to 1 in the past. It is unsatisfying to believe that we just happen to live at the time when Omega is just starting to depart from 1 by a small factor. It is much more appealing to consider that we do not live at a special epoch, so that Omega is still close to 1 today. But then we need to explain why Omega started out very close to 1 in the early universe. The theory of inflation provides just such a justification - most versions of inflation predict that the early universe was driven extremely close to flat, and that it is still very close to flat today. If this is so, then at least 90% the energy of the universe is dark! Note that although the universe may be flat that does not mean that matter makes up the critical density. In addition to dark matter there is dark energy, e.g. a cosmological constant that needs to be included in the accounting.

Stars of the New Aether Science

In the new scientific cosmology gravitational force has limited strength and plays less important role particularly on the movements of stars besides orbiting motions between stars and stars with their planets. Besides, the gravitational pressure on stars has nothing to do with producing nuclear reactions at the center of stars.

1. Formation of stars

In the beginning, stars can not be made of matters having atomic and molecular structures since they can not exist outside the universe. Baby stars were born having 100% dense-matter objects to begin with. Under constant bombardment of energetic atmospheric neutrinos, dense-matter objects inside the universe constantly release electrons, protons, neutrinos and huge amount of energy to quickly start nuclear reactions making hydrogen and helium. All stars undergo very long-lasting constant rate of nuclear reactions suggesting that their nuclear reactions only occur at their surface. Therefore, even atmospheric neutrinos can not penetrate deep into dense-matter objects, which have amazingly dense masses.

Therefore, the mystery of how stars can have nuclear-reaction at essentially constant rates for billions of years is now solved. All stars have a piece of dense-matter object at their center. Upon constant bombardment of the energetic atmospheric neutrinos, the surface of all dense-matter objects becomes radioactive constantly releasing protons, electrons, neutrinos along with a huge amount of energy to ignite and maintain a constant nuclear reaction rate likely for billions of years.

Stars constantly produce hydrogen, helium, and heavier elements, by consuming their dense-matter objects. Both the nuclear reaction rate and the size of a star are likely proportional to the the size of its dense-matter object. A star dies when its dense-matter object is nearly used up.

Dense-matter objects are essentially unknown to us. They are formed from the aggregation of protons and electrons outside the universe under very dilute low energy atmosphere of neutrinos. Inside the universe they are radioactive constantly undergoing nuclear reactions at their surface to produce hydrogen and helium. Dust and small size dense-matter objects inside the universe should have long been used up forming hydrogen and helium clouds in space. Small dense matter-objects have also been consumed to form mainly hydrogen and helium containing planets and even larger ones have formed stars. We can not even survive to get near a dense-matter object and have to rely on astronomical observations to verify their existence and to study them.

A large portion of the protons and the electrons freed from dense-matter objects by the bombardment of energetic neutrinos undergo nuclear reactions to form hydrogen and helium releasing some of their atmospheric neutrinos along with huge amount of energy. In this way the atmospheric neutrinos of the universe are continuously replenished at a constant rate by all stars. Upon constant releasing energetic particles, all dense-matter objects have antigravity property! Many dense-matter objects may have irregular shapes and, therefore, producing imbalanced pushing force in all directions by their own nuclear reactions. As a result, many dense-matter objects should have very strong rotational energy. This rotational energy together with nuclear reactions creates very strong electric and magnetic force fields and, thus, having a spiral wind of neutrinos around themselves which is likely the energy source keeping their planets in steady-state orbits as discussed earlier.

2. Features of Regular Stars

a. Having artificially low Density

A baby star was born to 100% dense-matter object and should be a strong antigravity star and invisible. It constantly makes hydrogen and helium throwing them out to a great distance to gradually form its outer layer. Stars therefore have a hot outer layer containing hydrogen, helium, and some heavier elements constant increase in their masses. This outer layer should have gravity by reducing the energy of atmospheric neutrinos entering it. As a star gets older, its outer-layer content gets higher; its gravity gets stronger while the antigravity of its dense-matter object gets weaker. Therefore, as a star ages, it gradually turns from an antigravity star to a gravity star. A mature star like our Sun has already turned into a gravity star. Since the outer layer of a regular star is at a great distance from its center or dense-matter object, it therefore appears artificially very large having artificially low density.

b. Developing explosive Property

Since dense-matter objects constantly throw out the hydrogen and the helium they make, it is therefore more likely that the outer layer of stars plays the role of making elements heavier than helium. As the outer layer of a star gets thicker, it, particularly, its inner surface gets hotter. Besides, it is constantly bombarded by energetic particles such as electrons, protons, hydrogen, helium, and neutrinos. It therefore will reach the critical temperature to begin nuclear reactions to make heavier elements. The outer layer of a young star is readily penetrable by protons, electrons, hydrogen, and helium. However, as its content of heavier elements gets higher, it gets harder and harder for these particles to penetrate through. Therefore, as a star gets older, its outer layer contains more elements heavier than helium and is harder for the gas particles constantly produced by its dense-matter object to penetrate through. As a result its inner gas pressure builds up. It may get to a point that a star can only periodically release its gas instead of continuously such as the release of solar wind by our Sun. Another factor is that the dense-matter object inside a star gets smaller all the time alone with its antigravity, thus, allowing its outer layer to gradually cool off and to reduce size. However, it continues to produce gas particles (hydrogen, helium, proton, electron), which continue to build up its inner pressure. A star therefore finally will explode and/or violently expand becoming a red giant, for example.

3. Dense Stars

a. Black Hole-like Dense-matter Objects

Huge dense-matter objects may be present particularly at and near the center locations of galaxies. Having both very strong anti-gravity property and rotation energy such a huge dense-matter object may have very strong spiral winds of neutrinos to blow all their hydrogen and helium out to a great distance to form a bright ring on a plain having a donut-shape structure leaving the dense-matter object at its center invisible such as the image obtained by Hubble Space Telescope in the center of galaxy NGC4261NASA.

b. Dense Stars obtained from Star Explosions

The findings of neutron stars, white and brown dwarf stars, and black holes offer strong scientific evidences to prove that dense-matter objects exists inside stars, which are either exposed or partially exposed upon the explosions of stars blowing all or part of their outer layer away.

Large stars undergo supernovae explosion to effectively remove their outer layer letting its dense-matter object exposed essentially invisible known as neutron stars and may be black holes. Their constantly release of huge amounts of energies proves that they are not dead stars. Instead, they become baby stars again! Their amazingly high mass densities show their high contents in dense-matter object. They constantly produce hydrogen and helium and throwing them out to a great distance to gradually form an outer layer becoming visible stars again.

Medium-size stars such as our Sun first violently expand to become red giant stars and then explode to shed most part of their outer layer leaving behind a small dense-matter object having a much smaller new outer layer, according to astronomical findings, containing mostly carbon and oxygen. Under the influence of its own gravitational force, the new outer layer is formed to have a minimum size balanced by the already greatly reduced antigravity force of its dense-matter object. Both the closeness of the new outer layer to its dense-matter object and the high content of elements heavier than helium make the new stars very small and hot. They are known to be white dwarfs. Findings show that white dwarf and even brown dwarf stars are very small stars having mass densities much larger than those of ordinary stars and ordinary matters having atomic and molecular structures. Findings also show that white dwarf stars are much hotter than ordinary stars. All findings prove that dwarf stars are high in dense-matter object content. Having high content of elements heavier than helium, the outer layer of a white dwarf is already not easily penetrable by the gas and the ion particles constantly produced by its dense-matter object. Being extremely hot, its inner surface layer should undergo nuclear reactions to produce even heavier elements faster than ordinary stars. As a result some white dwarfs violently explode again. They are known undergo type Ia supernova explosion, which is the most powerful of all the supernovae. Currently, scientists believe that white dwarfs are dead stars having no nuclear fuel (hydrogen) left. Actually, they are very active stars. The overall findings not only prove that white dwarf stars are not dead stars but also that there is a dense-matter object in all stars.

Stars of large dense-matter objects should burn nuclear reactions faster than those having smaller dense-matter objects to quickly accumulate heavier elements. As a result, they may have shorter lives and die by blowing up as supernovae to release their built-up pressure. Then, they become black hole-like or neutron star-like invisible baby stars again. After a while when they have accumulated thick enough outer layer of hydrogen and helium, they become visible stars again. Black holes and neutron stars are exposed dense-matter objects offering good opportunities to study them. Findings show that many neutron stars have very high rotational energies creating very strong electric and magnetic fields, etc.

Finding neutron stars, white dwarfs, and brown dwarfs having mass densities much higher than atomic matters proves the existence of dense-matter objects inside all stars as predicted by the new scientific cosmology. The brightness of the white dwarf stars and their supernovae explosions demonstrating that nuclear reactions are ongoing inside them and like neutron stars they are not dead stars neither. The dense-matter objects inside brown dwarf stars may be too small to sustain nuclear reaction on their surfaces but they are always radioactive to produce large amount of energy.

Although the dense-matter objects of neutron stars are bigger than those inside white dwarfs, neutron stars are not even bright enough to be directly observable. The findings may be explained by that the supernovas to produce neutron stars are more effective in shedding their outer layer materials than the process to produce white dwarfs. Neutron stars therefore may not have an outer layer or have only a very thin outer layer. Since dense-matter objects are invisible, so are neutron stars.

The dense-matter objects inside brown dwarfs are even smaller than those in white dwarfs. The dense-matter objects of larger brown dwarfs may be able to undergo nuclear reactions to produce hydrogen and helium. The dense-matter objects of small brown dwarfs may be too small to undergo nuclear reaction but are always radioactive constantly producing large amount of energy. Once their dense-matter objects have been used up, they become planets having low mass density containing mainly hydrogen and helium like Jupiter and Saturn.

According to the new scientific cosmology, the movements of galaxies and stars are mainly caused by the (spiral) winds of the neutrinos not gravity. Assuming that their motions are controlled by gravitational force alone, the mathematical models of modern physics requires the existence of very strong gravitational forces and much more masses than those can be seen in the universe to produce such strong gravitational forces in the universe. Modern physics therefore has to postulate that the universe mainly contains unknown dark matters. It reminds us that to fit the geocentric postulation, Ptolemy's mathematical model had to postulate that planets had to move in epicycles, etc. New cosmology has no need to assume the presence of either dark matters or dark energies to explain all universal phenomena.

Chapter 4 – Particle Physics

Electron and proton are the only two elemental particles serving as the building blocks of all atoms and their molecules. Neutron is only an electron-proton pair. However, physicists have found from both the debris of cosmic ray and high-energy collision experiments besides protons, electrons, and charged nuclei, other uncommon charged particles such as positron, muon, pion, etc. In fact, they have already found around a thousand of them in their high-energy particle-smashing experiments and they are building more and more powerful particle accelerators to find more.

Physicists' discovery of all uncommon charged particles is based on a single detection method: using electric field to detect their existence, what kind of charge they carry, their mass content, and the amount of charges they carry. They have found the common property of all of them: unlike electron and proton, they are very unstable. Actually, the existence of these particles has not even been confirmed by another independent method. For example, physicists have not been able to make and store any of them. Although charged particles should interact with one another to form nuclei, atoms and molecules like what electrons and protons do, no scientific evidence any of the hundreds of charged particles does. The non-existence of all these charged particles has been proven from the fact that they have not been found in any matters found on Earth.

The knowledge of the behaviors of charged particles made of electrons and/or protons in an electric field of electrons and protons have been well established. However, the behaviors of the non-electron non-proton charged or uncharged particles in the same electric field are completely unknown and unpredictable to us. The findings only suggest that electric field is likely a sensitive tool for finding charged particles. The findings of hundreds of unstable charged particles in the electric field may suggest that the high-energy collisions may produce many new AUPs, which may be temporary charged in the high energy environment.

What we have learnt from PM is that all phenomena are produced by matters having masses and relative motions and they must have energy source. So far scientists have recognized the facts that the so-called "charge" is some kind of energy. The existence of charged particles, their nuclei, atoms, molecules, are also universal phenomena requiring constant supply of energy to make them and to keep them stable. Ignoring the energy sources to produces all universal phenomena has been the common fatal mistake of the physical science community!

A. Modern Physics

The standard (mathematical) model of modern particle physics is based on the relativistic quantum theory of Dirac, which combined theories of relativity and quantum mechanics, of course, including all of their postulations and many new ones. To make Maxwell's theory of electrodynamics conforming to relativistic quantum theory, Dirac had led to the development of quantum electrodynamics (QED) postulating that electrons radiate electromagnetic waves (photon or light) when they are accelerated in an electromagnetic field. QED further postulated that photon was the particle that "transmits" the electromagnetic force and had further generalized to assume that every force was "transmitted" by certain particle. This has become the force-particle duality principle of modern physics. QED was the theory "successfully" predicted the presence of positron. The discovery of positron had "proven" its postulations that anti-matter existed, that all particles have their anti-matter particles counterparts, and that when particles collide with their anti-particles they annihilate converting 100% of their mass into energy obeying Einstein Equation.

Upon finding hundreds of "elemental particles", physicists have been forced to simplify their findings and offer some explanations. They therefore have classified particles into two main categories, the **Fermions**, those (electron, proton, neutrino, their anti-particles, etc.) made up matters, and the **Bosons**, which (pion, photon, their anti-particles, etc.) were postulated force transmitting particles. They also classified them according to particle size into **Baryon** (heavy particles of fermions, such as neutrons and protons), **Meson** (medium weight boson particles transmitting strong nuclear force) and **Leptons** (the lightest fermions including electron, muon, and neutrino particles). Also, **Hadron** refers to baryons and mesons together. In 1963 Murray Gellmann proposed that the properties of all known particles could be accounted for if the hadrons were made up of three more fundamental particles, which he called **quarks** (up, down, and strange). Since then, new particles were discovered that required the introduction of three more **quarks** (red, green, and blue). Since neutrons and protons are made up of quarks, the strong nuclear force between them must be a force between quarks, which is called glue and its force particle is called gluon. This quarks-gluons theory is known as quantum chromodynamics (QCD).

A renowned physicist described the particle physics of modern physics as "It is the experimentalist's nightmare: Everything has been predicted. Indeed, every time an accelerator is turned on, a new particle appears. Someone cracks that a Nobel Prize should be given for not discovering a new particle." Having so many unscientific postulations the standard model of the modern-physics-based particle physics is in fact a science fiction.

Scientists have been very successful in determining the relative masses of electron, proton, atomic nuclei, and molecular ions by Mass spectrometer using the electric-field method. The science of a proton-electron electric field has long been well established knowledge. However, applying it to non-electron and non-proton particles has not been proven correct at all! Yet, all the discoveries of the uncommon charged particles and the entire standard model of modern physics are based on this unproven application. Without considering "charge" as a special energy, physicists really know nothing about them. They need to have a fundamental understanding of PM that all universal phenomena are produced by objects having constant energy supply.

B. New Aether Science

Absolutely Uncharged Particles

The findings that high-energy particle accelerators can produce hundreds of subatomic particles suggest that the extremely high energy collisions inside stars would produce even more. It is likely that protons and electrons are smashed into their own absolutely uncharged particles (AUP) and even broken into smaller AUPs. Neutrinos may interact with some of them to form charged particles but some may remain uncharged like neutrinos themselves. However, we have conclusive evidence to show that all charged particles found besides protons and electrons are unstable and do not form or interact with one another to form stable matters since all matters found on Earth are made of proton and electron only.

It is likely that there are many AUPs like neutrinos. They are not a pair of charged particles like neutron. They are absolutely uncharged particles like neutrinos. If their sizes are close to those of neutrinos, we may call them neutrino-like particles. However, some of them may have much larger sizes than neutrinos. For example, the AUPs of electrons and protons should be much larger than neutrinos. They are those interact with neutrinos to form stable charged particles having their own atmospheres of neutrinos, thus, they do not exist alone as AUPs. The question is whether there are AUPs much larger than neutrinos but do not interact with them to form charged particles? If they do exist, they must have such unimaginably high mass density that nothing can effectively stop them. They may even be able to penetrate deep into and through dense-matter objects! All the experiments designed to detect neutrinos should also detect all AUPs, large or small! We are therefore not sure whether we have been detecting only AUPs or both AUPs and neutrinos since neutrinos may be too small to knock any atom off molecules to produce light. Fortunately, even if we have been detecting 100% AUPs, the amount of AUPs we have found is not a great threat to our very existence and even our health.

The Universe of Protons and Electrons

Neutrinos have turned out to be highly interactive particles responsible for producing all charged particles. However, only proton and electron form the matters we have found. Therefore, the most important part of the particle physics is still that of PM. The nature has its own very restrictive selection process, which has chosen only two particles, proton and electron, out of hundreds or even over a thousand detected charged particles the nature routinely produces in its very high energy environments such as the center of stars and high-energy laboratories to make all the matters in our world. Similarly, out of about one hundred of elements, the nature only choose mainly four (C, H, O, N) of them to construct organic molecules, which form plants and animals.

We have long found that protons repulse protons and electrons repulse electrons but protons attract to electrons. We therefore call protons positively charged and electron negatively charged. Since charge energy is due to the presence of an atmosphere of neutrinos surrounding an AUP and constantly maintained by the atmospheric neutrinos of the universe, there can be only one kind of charged particles. The same size charged particles may repulse each other with their own atmosphere of neutrinos. For two charged particles having very different sizes, there mutual gravitational force is stronger than their atmospheric repulsion force. As shown earlier, the above reasoning can explain and predict many close encounter situations of charged particles while the model of positive/negative charged particles has failed.

Our scientific knowledge and technology about electric charge, force, and field including magnetic force have been obtained from matters made of electrons and protons only. It may be possible that all charged particles are detectable inside an electric field of the protons/electrons system. However, the behaviors of charged particles other than those made of electrons and protons, especially those having unstable charges, in the electric field are unknown and unpredictable. For example, their masses and charges can not be quantified in the electric field of protons/electrons system. Although hundreds of charged particles have been detected this way, our knowledge about them, such as their masses, and amounts of charges, may be all wrong except that they are all unstable charged particles.

The producing of hundreds of unstable charge particles by high-energy particle-smashing experiments suggests that the AUPs of electrons and protons have been crashed to many smaller AUPs. Unlike the AUPs of proton and electron, the AUPs so produced are unstable charged particles and they appear in the electric field like ghost charged particles. We can conclude from the overall findings that only the AUPs of proton and electron have the unique property capable of continuously gaining energy from the atmospheric neutrinos of the universe to become stable charged particles.

Without sensitive, specific, and direct detection method to study AUPs such as neutrinos and those produced in high-energy smashing experiments, we still know essentially nothing about AUPs beyond what we known about neutrinos, which is very little.

The Universe of Neutrinos

The biggest surprise is that proton and electron are AUPs having their own atmospheres of neutrinos and therefore neutrinos are always the major component of all the matters made of atoms and molecules and only the charge properties or the properties of the neutrino atmospheres of electron and proton are important! The atmospheric neutrinos of the universe interact with the AUPs of electrons and protons to make them charged or to have an atmosphere of neutrinos of their own and continue to supply energy to keep them constantly charged and then further interact with the charged particles to produce electric and magnetic forces. They are the forces to put protons and electrons together to form atoms and molecules, the particles made all matters found on Earth. With their unusual stability, sizes, and abundance, electrons and protons appear to be the only nature-chosen charged particles to make the matters of this universe and even those in DON outside the universe since dense-matter objects are also made of protons and electrons. Since the charge properties of proton and electron belong to their atmosphere of neutrinos not their AUPs, pealing onion study to find out whether they are made of even smaller particles or not is likely fruitless. The particle-crashing experimental findings in high-energy laboratory have essentially confirmed this conclusion.

Nuclear Energy

Scientists have long discovered that there are minor weight differences among protons and may be among electrons too. They found that free protons and electrons are slightly heavier than those in the nuclei of light atoms. They also found that the protons and may be the electrons inside the nuclei of lighter elements are somewhat heavier than those inside the nuclei of heavier elements. They also found that nuclear fusion reactions involve losing small amount of masses. Theory of special relativity postulates that the mass lost in nuclear fusion reactions turns into huge amount of energy. The conversion of mass to energy is one of the many unscientific postulations of the theory, since pure energy is not even scientifically definable.

Nuclear fusion reactions therefore involve releasing some neutrinos from the atmospheres of protons and electrons. Rapid release of neutrinos from electrons and protons in nuclear reactions is likely the mechanism of producing large amounts of nuclear energies. In nuclear reactions, therefore, the energy released should be proportional to the amount of neutrino particles released by electrons and/or protons from their own atmospheres of neutrinos. Einstein Equation is only correct to show this proportionality and its constant may be close to but is not c^2.

The neutrino-content difference in both protons and electrons is therefore very important science relating to the major energy source of the universe. To continue supply the universe with both huge amount of neutrinos and energy, the protons and the electrons inside dense-matter objects should have the highest contents of neutrinos. Inside the universe, both free protons and electrons still have the highest neutrino content. Besides releasing neutrinos from nuclear reactions, dense-matter objects inside stars may also release neutrinos together with electrons and protons to produce amazingly huge amount of energy.

Chapter 5 –Problems in Physical Science

While scientists were debating what light was; wave or particle, they were also confused by that gravity force could act from distance. Then, they found electric and magnetic forces, which also appeared to act from distance. However, unlike the long hot debate over whether there was aether in space to produce and to transmit light, they have long developed and accepted no-aether postulations-based theories, such as Newton's universal theory of gravitation and Faraday's force-field concept, to interpret universal forces. Eventually, upon believing that they have proven the absence of aether in space, physicists had also given up their aether concept for light to embrace Einstein's theory of relativity timely developed in the early 20th Century. Theory of relativity is not only a no-aether theory but also greatly expanded Maxwell's pseudo-science believing that mathematics is the scientific origin of universal phenomena. Since both of its hypotheses have violated the fundamental teachings of PM, all no-aether theories particularly theory of relativity must be wrong and incompatible with PM.

A. Strong Signs of Serious Scientific Problems

That physics, besides PM, has serious scientific problems has been commonly acknowledged by many physicists themselves. For example, the flaws of modern physics and the cult-like behaviors of the physical science societies have been the subjects of several books, such as "The Final Theory" by M. McCutcheon, in 2002, Universal Publishers; "The Big Bang Never Happened", by Eric Lerner, Times Books in 1991; and "The Cult of the Big Bang, was there a bang" by William C. Mitchell, in 1995, Cosmic Sense Books. Besides, the following are a few examples of the many quotes exposing our lost physical science, direct from "physics experts":

"Gravity may not be working as advertised. Spacecraft hurtling through the Solar System have been behaving so bizarrely that some scientists wonder whether our theories of gravity are wrong." -- Charles Seife, New Scientist Magazine, September 1998

"We don't know anything. Everything about gravity is mysterious." -- Michael Martin Nieto, theoretical physicist, Los Alamos National Laboratory, in Discover Magazine, October 2003

"Only a few people understand -- or **think** they understand -- how a permanent magnet works. The magnet of everyday life is not a simple thing. It's a quantum-mechanics thing" -- Tatiana Makarova, physicist, Umea University, Sweden in Discover Magazine, December 2002

Although we hardly find opposing views in the Journals of physics, cosmology and astronomy, there is however an explosion of opinions and oppositions to modern physics in Internet discussion groups. For example:

Newsgroups: alt.sci.physics.new-theories, 6/27/2001: Why "The Einstein Hoax" (http://members.home.net/retiche/hoax.htm)

"The Einstein Hoax was not written because the author was on an ego trip, it was written because post-Einstein physicists have been degraded from a science which relies on the proper use of the rules of evidence and the rules of logic into a religion which is based upon the acceptance by the scientific community of items of **faith** as revealed by anointed **experts** but which have not been proven or may not even be provable. Once established, these **experts** have been insured the survival of that **faith** by controlling what material which gets to be published in scientific journals. .... The disturbing part of the present process is that the next generation of scientists is having their minds destroyed by being indoctrinated into the **true faith.**"

Another example given below is a casual Internet discussion on the author's posted gravity ideas, which interestingly pointed out the serious problems in the current gravity theories.

Newsgroups: alt. sci. physics. New theories, 12/26/1998: From JT

Anyone care to comment on a student-level solution to gravity? The level is so low that gravity is force or energy instead of (distorted) spacetime.

Matters are low potential or kinetic energy centers since things fall into them. Something, let us call it "space wind", must continuously flow into matters to create gravity. It must be particles with mass (too small to be detectable), capable of penetrating through matters, and traveling very fast, e.g., close to that of light. It must be such subatomic particles as neutrinos. Let us call it "space subatomic particles" or SSP.

A feasible mechanism is that as SSP enter large matters, such as earth, some collisions occur, resulting in their losing a fraction of their kinetic energy. Therefore, the large matters become low kinetic energy centers in space for SSP resulting in SSP wind continuously blowing into them to create and maintain gravity. The mass of a black hole is so dense that the entering SSP are captured by instead of penetrating through it resulting in maximum SSP wind or gravity.

An important side effect of this mechanism is that the SSP continue to lose kinetic energy in matters converting to heat energy. This newly identified heating energy is likely much larger than that produced by gravity pressure and is, therefore, the major heating energy source to ignite nuclear reactions in stars.

Therefore, the solution is that space is occupied by a dense atmosphere of SSP and the non-destructive interactions between SSP and matters produce gravity and heating energy to matters. It is also likely that SSP interact with matters, energized and non-energized, resulting in exchanging such energies as electromagnetic radiation and transmitting them through space.

This solution or theory conforms to the fundamental law of conservation of energy.

Participant 1: One comment – black holes result from near-infinite amounts of matter in one place creating near-infinite levels of gravity which break down atoms etc. With the type of gravity you put forward there will be a LIMIT to the amount of gravity a body can have dictated by the level of 'SSP' space wind available. Therefore, no black hole, no 'dark matter' is needed. But yes, it DOES provide a solution to the anomalous heating of Jupiter, for instance.

Participant 2: One comment – the phrase "near-infinite" is meaningless. Anything that is not infinite or indefinite is finite. The entire mass of the universe is, to our experience, a large number. However, any mathematician can quickly jot down a number that, if used to represent electrons, outmasses the entire universe by any arbitrarily large factor. Compared to this number of electrons, the mass of the universe is small.

Any useful understanding of physics requires a clear grasp of the concept of infinity. Newton himself could not develop the foundations of classical physics until he had developed the ability to cope mathematically with true infinites through the concept he called "fluxions." (For example, force is the derivative of momentum with respect to time.)

"Large" is a relative term, "infinity" is not. "Large" describes scalars, "infinity" does not. You can get "near" any given number. You can't get "near" infinity.

Participant 1: One comment – the phrase "near-infinite" is meaningless. Alright. Ne quibbleons-nous pas. The point was that if one holds that atoms themselves create gravity-energy then the more atoms, the more gravity. Theoretically this can then increase the amount of gravity to where it breaks down the atoms themselves. (This always brings a humorous image to my mind – that of a snake eating its own tail until "poof") If, however, gravity is attributed to absorption of some sort of 'space-wind' locally – then the greatest amount one can arrive at is when ALL the wind is blocked or absorbed. No more black holes to worry about. AND no more DARK MATTER needed to 'save' our current theory. A win-win situation!

Participant 2: This is well established. The first breakdown results in the fusion of hydrogen into deuterium and helium. The next stage, if there is sufficient mass present, is the fusion of helium into carbon and heavier elements. If there is a really big heap of mass (more than the sun has) then the atomic structure can break down entirely, and neutron matter is produced in HUGE amounts (this often happens catastrophically in a phenomenon called a nova). If there isn't enough mass to build a neutron star, then the star cools off to eventually become a dwarf star.

Why would a space wind effect be proportional to the mass of a body and not its cross-sectional area? How could a space wind produce a conservative spatial potential field? How could a space wind effect produce a gravity-like force on a body without creating a friction-like drag on moving body? Answer these analytically and I will listen to you. Convenience in preserving a cherished concept (our current theory) or in avoiding a concept some may find awkward (no more black holes to worry about) are NOT reasonable arguments to support a scientific theory.

Participant 1: What IS reasonable is this: when talking space-wind, i.e. incoming energy of some sort that produces gravity, we already FAR ahead of current gravity theory.

Consider – a planet with a mass a certain distance away. The mass falls into the planet (via gravity). Now we must do work, e.g. spend energy, to return it to its formal place. We let it go. It falls again. We do yet MORE work. Hey, what's going on? It appears we could do work indefinitely returning this mass, spending huge amounts of fuel, and the planet just sits there unchanged? "Where is it getting the energy to move the mass? If the energy is being spent to move it one way, it must be spent to move it the other. WHERE is it coming from??? Matter bends spacetime? Say what? Why? How? Hunh? Characteristics of the field, drag effects, etc., hell, at least we take it from there. Look at Participant 4's stuff. He has some explanations for your questions.

Participant 3: Any theory relating gravity as a pushing force results in a universe totally at odds with that that created us and which we observe today. The most demonstrative proof of this would be to climb into a thick lead sphere. Inside there should be no gravity as you would be completely shielded. If as you postulate, the particles are capable of penetrating all matter, then they can exert no force, hence we would all be floating in a cosmic sea with no light, there being no stars to generate light and no planets to smack into. Since none of this is so, your theory isn't either.

Participant 4: That created us? The most demonstrative of this would be to climb into a thick lead sphere. Inside there should be no gravity as you would be completely shielded.

Ha ha ha, ROTFLMAO... OK, try neutrino (as just one example), how thick must your lead be to stop them?

If as you postulate, the particles are capable of penetrating all matter, then they can not exert no force,

Really, again just as an example, do neutrinos pass through matter TOTALLY unattenuated? If not, they WILL exert a force...

Hence we would all be floating in a cosmic sea with no light, there being no stars to generate light and no planets to smack into, since none of this is so, your theory isn't either.

Not his theory son, Le Sage's, circa 1780's... Try again when you have an inking of subject your attempting to talk about. Taking a stance from a position of ignorance is the mark of a fool.

B. The Geocentric Era - The Wrong Beginning

Physics had a trouble beginning. Ptolemy was an astronomer and mathematician. He believed that the Earth was the center of the Universe so was the powerful church at the time. The word for Earth in Greek is geo, so we call this idea a "geocentric" theory. Even starting with this incorrect theory, he was able to combine what he saw of the stars' movements with mathematics and geometry, to predict the movements of the planets. In order to make his predictions correct, he worked out that the planets had to move in epicycles, smaller circles, and the Earth itself moved along an equant. None of this was true, but it made the mathematics work for his predictions. This flawed view of the Universe was accepted for well over a thousand years. Apparently, he had started the earliest religious pseudo-science in astronomy using a mathematical model based on religious beliefs or unscientific postulations and the support of very limited amount of astronomic observations. It was a religious pseudo-science era since correct scientific views, such as those of Copernicus, were forbidden by the authority, the powerful church at the time.

Copernicus (16th Century) came up with a new way of looking at the Universe. His heliocentric view put the Sun at the center of the solar system. He used his own observations of the movements of the planets to back up his idea, which were however too revolutionary to be acceptable (even for the public) at the time. The powerful authority against this new scientific idea was the church. He and his followers, the Copernicans, therefore, had to communicate their scientific ideas in secrecy. About a century later Galileo was able to build a telescope and used it to prove that Copernican's heliocentric view was correct. He published his observations which went against the established teaching of the Church. He was brought to trial and, although he made a confession of wrong-doing, he was still imprisoned for life. However, his publication had finally helped ending the era of the religious pseudo-science. Besides, the physics of mechanics (PM) he has led to develop in the 17th Century has been repeated proven to be the only real science, the only correct physics and the fundamental science for all fields of science ever since.

C. The Rise of No-Aether Theories

Although the heliocentric view of Copernicus had been accepted, the scientific revolution in cosmology was only the beginning. The successful development of PM unfortunately had not stopped the mistakes scientists had continued to make in both physics and cosmology. Long-time failure in finding the aether predicted from the teachings of PM to produce light and universal forces had eventually made physicists to betray PM by developing no-aether postulations-based theories. In earlier days all scientific evidences such as planetary motions pointed to static universe strongly suggesting the presence of nothing (no aether) in space to slow down motions. The no-aether gravity concept was developed in 18th Century by Newton, the famous successor of Galileo having important contribution to the development of PM. The phenomenal success in his universal theory of gravitation had led to the acceptance of other no-aether postulations-based theories leading to a new era of religious pseudo-physical science lasted for a century already. A brief review of all no-aether theories and their flaws are given below.

Newton's Theory of Universal Gravitation

Physicists have long ignored the fundamental science of PM leading to the accumulation of the biggest scientific mistakes over several centuries. The strangest thing is that the scientific mistakes began from Newton, one of the famous founders of PM, in the 18th Century. His famous universal theory of gravitation postulated that all objects had act-from-distance attraction force to produce gravity. These postulations violated the teachings of PM since PM only teaches the presence of act-on-contact pushing force of objects. The presence of attraction force having the ability of act-from-distance contradicts the teachings of PM. Besides, according to the law of conservation of energy of PM, to continuous produce force, such as gravitational force, a source to continuously supply energy is needed. Scientists still have not found the energy source that enables all matters to continuously produce their gravitational forces.

The mathematical principle of Newton's gravity theory is: gravitational force between two objects is proportional to the mass contents of the two and inversely proportional to the square of their distance. It is correct for calculating and predicting the orbital motions of the planets in the solar system as proven by many observations and astronomical findings. As a result of having wrong science, however, its mathematics is only partially correct. For example, scientifically, there should be no infinitely strong gravity force. Also, there is no universal gravitational constant and therefore it is wrong to use the constant for planets to stars such as our Sun. There are many scientific factors affecting the strength of stars' gravitational force, such as their age, temperature, their production of neutrino particles, etc, which have not been taking into consideration by Newton's theory.

Accepting Newton's universal theory of gravitation has also accepted its unscientific postulations. It opened the door to no-aether postulations based theories. Also, the main reason for scientists to accept his theory was that its mathematics could predict the orbital motions of the planets in the solar system. The phenomenal success of Newton's gravity theory through mathematics may have misled physicists to begin to use mathematics to solve scientific questions.

Faraday's Force Field Theory

Upon observing iron files forming lines around the poles of magnets, Faraday developed "force field concept" in 19th Century, to explain why universal forces could act from distance. He visualized the space filled with lines of force that told an object what direction to move and with what acceleration. Therefore, his force field is an "object-free medium" to give strength and directional "messages" of force, which does not make any sense scientifically. However, mathematically it is useful for offering convenience in calculations of a force at any location. Force field theory has long been generally accepted for the interpretation of all universal forces such as electric, magnetic, and gravitational forces. The postulation of the presence of any kind of energy (force fields) in space without having objects to produce it is against the teaching of PM, thus, unscientific. There are scientific force fields, which are particle fields as described in Chapter 2.

Maxwell's Light Theory

Using mathematics to interpret Faraday's postulated electric and magnetic force fields, Maxwell in 1860's had derived a mathematical equation known as electromagnetic wave. He surprisingly found that the mathematical "propagation speed" of this wave function was the same as the light speed found experimentally. He therefore postulated that light was electromagnetic wave. In 1888 Hertz discovered radio wave, an invisible light, by discharging electrons between two wires having electrical potential difference. Physicists believed that Hertz had proven Maxwell's theory that light is electromagnetic wave. Even today, we all firmly believe that light is electromagnetic wave, a mathematical function!

Even today, physicists still do not know what a force field is. If electromagnetic wave has any scientific meaning, it should be somehow related to the force fields assumed to exist. To prove that light is electromagnetic wave; first we need to know what the electric and magnetic force fields are. The valuable discovery of Hertz proved that the oscillation of electrons (between two wires having different electric potentials) produced radio waves, an invisible light. He has not discovered what light is, what force fields are, and certainly has not proved that light is electromagnetic wave. According to PM, light is a phenomenon, which has energy and must be produced by some kind of object(s). For example, sound is produce by air particles. To prove that electromagnetic wave is light, we must also prove that a mathematical function can be an object, which has mass and relative motions. Without having unique physical and chemical properties, mass, occupying space, and having relative motions, a mathematical function can not be possibly representing or being an object. It happens often that two unrelated objects travel at the same speed, which is just a coincident and cannot be use by science to prove that they are the same thing.

Accepting Maxwell's postulation that light is a mathematical function has far-reaching consequences. It marked the beginning of turning physics into a religion believing that mathematics is the Lord of nature science.

Modern Physics

Modern physics mainly includes theory of relativity, quantum theory (mechanics), and big bang theory. It accepts force field theory and Maxwell's light theory with minor conceptual modifications. Besides having "no aether" postulations, it has greatly expanded Maxwell's "mathematics-derived pseudo-science". In the past century Einstein's scientific revolution has turned the entire physical science into a religious-belief and mathematical-model based pseudo-science amazingly similar to that of Ptolemy.

1. Theory of Special Relativity

Its fundamental postulations that light can travel through space by itself without the need of a transmission medium are unscientific according to PM and must be wrong. Assuming that light traveling at an absolute constant speed in space also violates the fundamental science of PM. PM teaches that all objects have relative motions. Since phenomena are produced by objects, they must also have relative motions. Having absolute constant speed means that light does not have relative motions like all other objects and phenomena. It is scientifically impossible for any object and phenomenon to do so.

Under the assumption that light travels at absolute constant speed, mathematics can prove that both the time and the size of objects are not absolute but functions of their speed and the mass of the environment. For example, the mass of an object would increase with its traveling speed and would reach infinitely large when it reaches light speed. In high-energy particle laboratories, physicists do find that the faster a particle travels the harder to further accelerate it and they are unable to accelerate any particle such as electron to light speed. They religiously believe that they have proven Einstein's theory of special relativity. Actually, what they have found is the friction of aether to the fast moving particles they have been accelerating. They actually have found strong scientific evidence for the presence of aether. How could they ignore such simple scientific explanation they encounter daily, for example, the presence of the resistance of wind upon riding a bicycle or running to choose some unscientific postulations-based interpretation?

Another major effort to prove theory of special relativity is to prove that time is not absolute. A lot of physicists believe that they have enough scientific evidences to prove so! An experiment often cited to prove that time slows down with increasing traveling speed was that a spaceship circulating Earth at high speed carrying an atomic clock synchronized with another atomic clock on Earth. The finding was that the clock on the spaceship was very slightly slower than that on the Earth. If this finding is valid, it would be due to the difference in the nuclear decay rate between these two atomic clocks due to the difference in the atmospheric pressure of neutrinos at these two locations. According to the new aether science, the energy level of the neutrino particles in the surrounding space (onboard satellite) is higher than that on the surface of Earth. The atomic clock onboard the satellite would be expected to be slower than that on Earth.

Of course, we must ask whether the famous Einstein Equation $E = mc^2$ is scientifically correct or not. The answer is that it can not be correct. Modern physics postulated that mass (object) could be converted to (pure) energy obeying Einstein Equation. According to PM only objects produce energies and no energy exists in pure form or in the absence of any object. The term "pure energy" is scientifically indefinable in the absence of any object. Even modern physics has not come up with a clear scientific definition for "pure energy". It means that Einstein Equation contains a scientifically indefinable term, thus, does not have any scientific meaning.

We have learnt from modern physics that in all nuclear reactions a small portion of mass is lost turning into large amounts of energy obeying Einstein Equation. This interpretation is of course based on the postulation that mass can be converted to pure energy and it cannot be scientifically correct unless both that "pure energy" is scientifically definable and that PM is proven wrong. We also have known for some time that nuclear reactions release neutrino particles, which have been found to have small mass in several recent findings. It is therefore also possible that the mass lost in nuclear reactions are due to the release of neutrino, which are too small to be directly detectable.

The question remains in the above scientific interpretation is that where the extremely large amounts of energy released by nuclear reactions comes from? Like nuclear reactions, chemical reactions also often release different amounts of (heat) energy even though they are much smaller than those of nuclear reactions. The energy released is the difference between the total chemical structural energies of the reactants and those of the reaction products. All chemical reactions obey the Mass-Balance Law of PM so are all physical reactions or changes. There is no scientific reason why the nuclear reactions do not obey the Mass-Balance Law. Like chemical reactions, physical reactions also often release or absorb heat energies. However, the energies in physical changes are much smaller than those in chemical reactions due to the big differences in their structural energies. Physicists have failed to utilize the well-established scientific knowledge of chemical and physical reactions to understand nuclear reactions. Assuming that the energy of nuclear reactions comes from the conversion of mass into energy again violates the teachings of PM.

Atoms have both nuclear and atomic structural energies. The energies released (or absorbed) by nuclear reactions should be the total difference in structural energies between reactants and products like chemical reactions. To calculate the energy released in a nuclear reaction, we need to find out the structural energies of the atoms of all the reactants and all the reaction products and take the difference between them. Instead physicists use Einstein Equation to calculate nuclear reaction energy. There is no scientific reason to believe that these two results would agree with each other.

A nuclear reaction uniquely involves simultaneous releasing of many kinds of energies, such as heat, radiation, the energies of the releasing subatomic particles, etc. It is very doubtful that all the energies released by nuclear reactions can be quantified with accuracy experimentally. Einstein Equation is just one of the religious beliefs of modern physics instead of real science. Experimental findings show that the amount of energy released by a nuclear reaction is proportional to the amount of mass lost or the amount of neutrino particles released. A science interpretation of the energy-releasing mechanism in nuclear reactions is given in Chapter 2.

Theory of special relativity also postulated that light is pure (quantized) energy known as photon. Since pure energy is indefinable, this postulation again does not make any scientific sense and is another violation of PM.

2. Theory of General Relativity

Theory of general relativity is Einstein's theory of gravitation having very complicated mathematics. However, it is not necessary to understand its mathematics to scientifically judge it. First of all, to be scientifically correct it must be consistent with PM, which is the only proven correct science in physics. Besides, upon the discovery of aether, the PM-based science of gravitation has already been discovered, which disproved theory of general relativity. Nevertheless, it is a very imaginative religious pseudo-science having fooled the entire world for a century already.

Science aside, Einstein had introduced this theory by making some miracle-like predictions such as that star lights passing through the surface of Sun should be bent by its strong gravity, which have been proven to be correct! The entire world has been shaken by his miraculously successful predictions, which has raised the status of Einstein and his theories of relativity beyond science. As a result, his "scientific revolution" has enjoyed an instant success, thus, creating another era of "religious pseudo-science" in the 20th Century since the ending of the geocentric era in the 17th Century!

Again, theory of general relativity is a no-aether theory. One of its important postulations is that gravity is equivalent to acceleration. This is only a general statement that all forces produce acceleration or is equivalent to acceleration. It is useless for finding what (object) produces gravity. Theory of general relativity uses (force) field equations to calculate gravitational force and to interpret the expansion of the universe. Since both time and dimensions were not absolute but functions of moving speed and surrounding masses, theory of general relativity introduced four-dimensional spacetime to replace the conventional three dimensional space and absolute time. Einstein further postulated that spacetime have geographic features: "The geography of the spacetime is determined by the distribution of the masses and the energies in the universe. In other words, the masses and the energies tell spacetime how to curve." He further postulated that curved spacetime produces gravity, which suddenly is not a real force any more! Even today most physicists still can not really accept this concept. After all, theory of general relativity is based on Einstein's postulation that gravity is equivalent to acceleration and only forces make objects to accelerate!

The phenomenal success of Einstein's theory of general relativity was due to the similarities of its 4-dimensional spacetime universe to the predicted aether universe of PM. Upon putting in the empty space something (spacetime) that can (interact and) be distorted by matters and their energy, theory of general relativity actually mimics a having aether theory! These similarities had enabled him to make some predictions, which would be expected in a having-aether universe but are miracle-like in a no-aether universe. For example, Einstein had boldly predicted some having-aether gravity-related phenomena, such as light-bending and gravitational red-shift. The amazing similarity of the curvature of spacetime to variation of the energy level of an atmosphere of aether (neutrino particles) allows theory of general relativity to calculate the planetary motions of the solar system more accurately than that of Newton's theory of universal gravitation, whose calculation approximates stars and planets as points. One well-known example is that Einstein offered more accurate calculation on the axial rotation of Mercury in its elliptical orbit at the point closest to the Sun. This is however only mathematical improvement not relating to the correctness of the science of these two theories.

According to theory of general relativity there were time differences in different locations inside a region having curved spacetime, which was postulated to be the scientific origin of gravity and gravitational red-shift, etc. The new aether science however attributes these phenomena to the kinetic energy variation of the atmospheric neutrinos. Since both scientific interpretations are incompatible, one of them must be wrong. The one which is compatible with PM should be the correct science.

Recently physicists are publicizing using Global Positioning System (GPS) to prove that time is not absolute. They claim that GPS has building in "time correction" for the effect of both theories of special relativity and general relativity in order to work correctly. According to them: because the moving velocity of GPS satellites, onboard clocks run about seven microseconds slower per day than ground clocks. The weaker gravitational pull on the satellites adds another relativistic effect, making them run 45 microseconds faster per day. Hence, a correction factor must be calculated that effectively turns back onboard clocks by 38 microseconds per day to yield accurate GPS data. The correction needs to offset the "frequency in satellite". Many physicists including an officer of NSF claimed that "every time GPS accurately locates a position also proved both theories of relativity were correct".

Again, the frequency difference between the satellite location and that on Earth is predicted by the new aether science, which would also show up as time difference by atomic clocks. However, the new aether science predicts that an atomic clock goes slower onboard a satellite in space than on earth irregardless of the moving speed of the satellite. This prediction contradicts that of Einstein. The author challenges physicists and NASA to determine which is correct.

It is also time to return to common sense. In science, simpler answer is preferred. The one without postulation or with fewer postulations should be the choice. The above scientific criteria should choose the 3-dimensional universe of the new aether science over the 4-dimensinal universe. When we can explain universal phenomena better with the atmospheric neutrino particles as the aether, why we need no-aether postulations-based theories? Why cannot we apply the well-established knowledge of sound and the entire PM to understand light? Why we continue to insist on that light does not have a transmission medium after that aether has been discovered?

We have mastered the science of the global phenomena for a long time already. We know that the presence of an atmosphere of air allows worldwide phenomena such as wind, sound, fire, etc. to occur anywhere inside it. For all universal phenomena, such as light and universal forces, to occur anywhere in the universe there should be an atmosphere of aether to produce them. Instead of applying the well-established knowledge of worldwide phenomena to understand universal phenomena, what Einstein and physicists have been doing is to use no-aether postulations to derive mathematical functions or mathematical models to be the scientific origins of each universal phenomenon, such as light and gravitational force. Then, they develop a "pseudo-science" to interpret them. It would like that instead of scientifically discovering the existence of an atmosphere of air to produce global phenomena, physicists had postulated nothing in space to produce global phenomena. They then derived mathematical functions to be the scientific origins of wind, sound, fire, etc. and develop a mathematics-based "pseudo-science" to interpret each of them. This way we could never discover and understand the science of air and its global phenomena. It has been a century of modern physics already, when we have learnt a lot about light, gravity, and the universe. Yet, nobody really knows what they are besides some unscientific and religious beliefs, which only add to our misunderstanding of the nature.

3. Big Bang Theory

Big bang theory proposed that the universe was once extremely compact, dense, and hot. Some original event, a cosmic explosion called the big bang, occurred about 10 to 20 billion years ago, and the universe has since been expanding and cooling.

The theory is based on the mathematical equations, known as the field equations, of the general theory of relativity set forth in 1915 by Albert Einstein. In 1922 Russian physicist Alexander Friedmann provided a set of solutions to the field equations. These solutions have served as the framework for much of the current theoretical work on the big bang theory. In 1927, the Belgian priest Georges Lemaître was the first to propose that the universe began with the explosion of a primeval atom, based on the findings of the red-shift in distant nebulas by astronomers. American astronomer Edwin Hubble provided some of the greatest supporting evidence for the theory with his 1929 discovery that the light of distant galaxies was universally shifted toward the red end of the spectrum. These findings, under no-aether postulations, proved that the galaxies were moving away from us and from each other. Hubble found that galaxies farther away were moving away faster, showing that the universe is expanding uniformly. However, the universe's initial state was still unknown. In the 1940s Russian American physicist George Gamow worked out a theory that fit with Friedmann's solutions in which the universe expanded from a hot, dense state. In 1950 British astronomer Fred Hoyle, in support of his own opposing steady-state theory, referred to Gamow's theory as a mere "big bang," but the name stuck. The theory essentially suggested that the universe began with a big bang of something called singularity having infinitely dense mass, high temperature, and strong gravity and was uniformly expanding ever since.

Originally, the field equations of the theory of general relativity would allow the universe to expand or to contract. Like everybody else, Einstein was convinced at the time that the universe was static. He therefore added an extra, ad hoc cosmological term to his equations to counterbalance gravity, thus, allowing a static solution. After he had learnt Hubble's findings and the conclusion that "the universe was uniformly expanding", he had abandoned the cosmological term to allow the universe to expand as a solution and admitted that the introduction of the cosmological term was the biggest blunder he ever made in his life. Now, the big bang theory has been included as part of theory of general relativity by many physicists.

Redshift is a Doppler Effect. The science of Doppler Effect in sound is well established. A Doppler Effect may be caused either by relative motion or by transmission effect. Upon postulating "no aether" to transmit light, theories of relativity and its modern physics have eliminated the entire scientific possibility of transmission medium effects leaving the only scientific conclusion that the red-shifts found in star and galaxy lights must be due to relative motions and the universe is uniformly expanding. Now, there are strong astronomical findings to prove the presence of aether to transmit light and that the universal redshift of star lights is caused by the medium effect. Wrong scientific conclusion has misled our modern cosmology for a century already!

4. Quantum Theory

Max Planck had led to develop Quantum theory in 1900 in an effort to explain why the spectra of all black bodies are the same, which only change with temperature. The essence of the theory is that certain (energy) quantities in nature occur only in discrete intervals; the size of these intervals can be determined by Planck constant h. The spectra of black body can be explained using his equation $E_q = n\ h\ \upsilon$, where E_q is quantized energy; n must be an integer; h is Planck constant, and υ is a frequency of a black body spectrum. Niels Bohr had successfully applied the quantum concept of Planck to the Rutherford's atomic structure of hydrogen to explain its line spectrum. He postulated that the only allowed orbits of the hydrogen's electron are those in which the angular momentum of the electron is quantized and that only when electron jumps from one orbit to another to radiate or absorb light. Erwin Schrödinger and Werner Heisenberg had independently developed mathematics to calculate the probability of electrons in the quantized orbital for essentially all atoms and molecules leading to the understanding of their line spectra and their atomic and molecular structures. These are great scientific accomplishments of quantum theory. The Uncertainty Principle of Heisenberg "it is impossible to measure both position and momentum of a particle with arbitrary precision" is also worth mention, which defines the limitations of experiments.

The above-mentioned is now called classic quantum theory. It is however the only part of the quantum theory, which is largely scientifically correct. Planck's quantum principle apparently was developed for "collective energies" such as the energies of all electrons in atoms and molecules. Quantum theory is however scientifically wrong to be expanded as a general scientific principle! This is because that not all collective energy is quantized and quantum theory is applicable only to those energies of electrons in atoms and molecules for special scientific reasons. It is like the sound of piano, which are quantized for a more obvious reason. For example, forces are energies but they are not quantized. The fundamental science of PM teaches that all energies are produced by objects. It means that at the level of a single particle, energy is always quantized. For example, at the level of single electron, the energy of the electron, either orbital or free electron, is quantized. This general scientific principle, however, can not be used to explain the phenomena of certain collective energies which are quantized for some special scientific reasons. Other major scientific problems of quantum theory are that it honors the no-aether postulations of theories of relativity. The above-mentioned part of quantum theory however is largely not affected by those postulations besides limiting its overall scientific understanding.

The combination of theories of relativity and quantum theory however became a religious pseudo-science famous for its weirdness. Dirac's relativistic quantum mechanics (RQM) postulated the presence of anti-matter particles. Upon finding positron, physicists had claimed that RQM had been proven correct and was again generalized to claim that all particles have anti-particles. Positron was one of the several hundred ghost particles found in high-energy particle crushing experiments and cosmic radiations. Physicists however have not been able to isolate and to prove their existence in any matters. RQM further postulated that, as all Star Trek fans know, when a particle and an antiparticle meet, they annihilate, converting 100% of their mass into energy! The theory of quantum electrodynamics (QED) Dirac led to develop postulated that all universal forces were transmitted by special particles. The weird part is that some of the force-transmitting particles have masses and some don't! For example, those transmitting light (photon) and gravity (graviton) have no mass. Yet those transmitting strong nuclear force (pion) and weak force (w) have masses. Modern quantum mechanics further postulates that the space is not really empty and it has energy, a form of energy that arises from the so-called virtual-particle-antiparticle pairs. Physicists postulated that there are virtual particles-antiparticles pop out of the vacuum, travel for short distances and then disappear again on timescales so fleeting that one cannot observe them directly. Modern quantum theory is a science fiction and its development has been led by science fictions too.

Incompatibility with PM

PM is the having-aether science that must be completely incompatible with all no-aether theories. However, physicists believe and even insist in that modern physics is compatible with PM. They have even included PM as part of modern physics. According to them PM has a valid range. Theory of relativity must replace PM when there are objects moving at relativistic speed. Besides, in a micro environment energies are quantized and quantum theory must be used to replace PM.

A proven correct science like PM should be "the science of everything" and in the absence of special scientific reason should have no range limitation. Only wrong science and wrong theories are explainable only by some experimental findings, thus, appear having valid ranges. A wrong theory can never be "a theory of everything". Therefore, the dream of Einstein and many physicists to have a theory of everything can never come true.

PM is a having-aether science, which is completely incompatible to all no-aether theories. There is neither a need to make relativistic velocity correction nor that all the energies in micro-environment are quantized.

D. The New Era of Religious Pseudo-Science

Throughout the 19th Century most scientists believed the presence of aether in space to transmit light. However, physicists continued to be frustrated by their failure in finding aether. By the end of the 19th Century many of them had turned to no-aether postulations-based theories. Some no-aether theories such as Newton's universal theory of gravitation and Faraday's force field theory had already been accepted by the scientific community. Even Maxwell's light theory, the mathematically derived pseudo-science, had gained acceptance. Meanwhile, scientists still believing aether concept faced unanswered aether questions and did not know what to do.

At the beginning of 20th Century, Einstein timely introduced his theory of relativity with bold predictions. After his prediction that light passing through the surface of the Sun should be bent at an angle as he calculated by Sun's strong gravity was confirmed correct in 1919, Einstein has become famous worldwide not only as a scientist but a symbol of wisdom and more. His theory of relativity was instantly hailed to be the biggest scientific discovery ever. It was accepted worldwide with the support of only very few experimental findings, which only proved some of its mathematics but not its science! For example, experimental observations proved that Einstein was correct in both predicting and calculating the warping of light around the Sun. However, the findings have not scientifically determined whether the warping of light was due to curved spacetime, the attraction force of the Sun (Newton's theory), or the presence of aether. Blindly believing Einstein's theory of relativity by both the scientific community and the public has led to a new era of religious pseudo-science.

What Einstein had started was a new religion by expanding the already accepted no-aether theories, their beliefs, and the mathematical pseudo-science of Maxwell. He timely offered a "solution" to the aether issues the scientists were desperately seeking and they therefore had enthusiastically taken it. Einstein's theories of relativity openly defied the common wisdom or the common sense based on the scientific knowledge of PM. Yet, it was quickly accepted by the scientific community since many no-aether theories had already been accepted. Of course, there were strong oppositions, for example, by German scientists. They might have stopped Einstein, if they had won the war.

Einstein had started a (anti-)scientific revolution by disregarding accumulated scientific findings and knowledge and by crashing down oppositions enforcing its own beliefs governing its scientific communities with iron fist. It is an anti-science revolution against PM, in particular, its teachings that there should be aether (object(s)) in space to produce universal phenomena. It has quickly succeeded in the fields of physics and cosmology. Even a century later, however, it has not affected all other fields of science due to its apparent inapplicability in them. Its inapplicability in so many fields of science, which continue to base on the science of PM (objects based), clearly indicates that no-aether theories are wrong.

Modern physics still requires experimental findings to prove the correctness of its mathematically derived functions to be the scientific origins of universal phenomena, since both scientific communities and the public would never give up this requirement of PM on science. Physicists however have managed to confuse the public with experimental proofs of mathematical calculations as scientific proofs! Nevertheless, this requirement has severely restricted the progress of modern physics, since not many wrong sciences are accidentally supported by some experimental findings. For example, it is not often for such coincidences to happen as that electromagnetic wave has the speed of light and as that curved spacetime mimicking the presence of aether well. In fact, it has been so hard to find such coincidences that no significant progress has been made for modern physics in the past half century or longer, despite that billions of tax dollars have been spent together with the concentrated efforts of the entire physical science society. As a matter of fact no real scientific progress has been made in the past century. What physicists have been doing is to religiously expand the influence of modern physics. They have been busy to fool the public, to control and to rule the physical science society with iron fist by establishing strict rules and faulty scientific criteria to protect their theories and to deny PM-based concepts and findings. They have been very successful in controlling the scientific funding worldwide to further modern physics research and education essentially unchallenged in the past century!

Modern physics is based on religion-like beliefs or unscientific postulations ignoring scientific understanding. It mainly preaches a mathematically derived pseudo-science of the universal phenomena based on postulations of no-aether (no objects) to produce universal forces, light, etc. Its (scientific) method is based on its formal science philosophy requiring the development of mathematical models or theories to discover universal phenomena. For example, discovery of an object (aether) experimentally to be the scientific origin of a universal phenomenon is not acceptable without a mathematical theory to back it up, which is impossible anyway. In the universe of modern physics people believe that mathematics is the Lord of science and all universal phenomena and even the universe are mathematical functions. The requirement of mathematical theories to back up discoveries has been a criterion particularly designed and used to deny PM-based real science or having-aether science. Together with the use of the unscientific beliefs-based knowledge of modern physics as scientific criteria the authorities have very effectively blocked PM-based ideas and discoveries including the new aether science from publishing in scientific journals and in getting research funding.

In the era of the religious pseudo-physical science only modern physics based interpretations of findings and ideas are officially acceptable for publication and for getting research funding. The majority of the members of the physical-science societies are professors, teachers and researchers of universities, colleges and schools and they are the stubborn supporters of modern physics. Children are taught modern physics concepts in schools and at home starting at very young age and they continue to learn them throughout schooling ages. The public is also bombarded daily by modern physics in news, science magazines, books in public libraries, and even many science fictions are modern physics based. Modern physics has long become part of our modern culture!

Both aether and aether science have finally been discovered recently by the author. Normally, scientific discoveries, particularly of great importance, are welcome and published in scientific journals. Unfortunately, it has been over eight years since the discoveries and the author has exhausted all his effort to do so without success. It's been centuries that no-aether-theory and mathematics-model based physics has been used to interpret gravity, electric, magnetic forces and light and a century already since modern physics has become the only officially acceptable physical science. The views and even the findings that aether exists have long been officially unacceptable! Authorities such as those of NSF, physical science societies, and scientific journals have long turned their back to real science!

The history therefore has repeated itself that the authority of science has become the enemy of science. This time it is not the church but the very scientific and educational organizations established by the governments to promote, educate and protect science! However, this time the public basically understand what science is and, in the presence of new aether science, should not be fooled for very long.

Intelligent Design

Another major battle ground between science and religion has been evolution versus creation. The Supreme Court of the United States has ruled favoring evolution. The ruling has also sided PM since although evolution is a largely unproven theory, it is consistent with PM knowledge while creation is not. Recently, the battle heated up again in the field of biological science. Many scientists are challenging the evolution theory of Darwin, which remains largely unproved. Challengers however have not offered a better scientific alternative beyond creation and intelligent design. Creation has long been generally regarded as a religious belief unacceptable to scientific community. Intelligent designs may be a theory to explain unexplainable but scientists would still want to find out how the designs work. The relevance here is that modern physics, such as its teaching about big bang theory and planetary motions, has often been used to support intelligent design and even creation arguments.

Unlike religion, scientific knowledge is limited to what we know so far. Science is part of human and therefore is often wrong too. The spirit of science is however keeping trying until the correct answer is found. A real scientist would not take intelligent design as a final scientific answer and believes that there is a scientific reason behind everything and every phenomenon. For example, scientists have not given up finding the reason(s) behind the centuries-old findings that planets in the solar system can stay in their orbits for god knows how long already! Many may believe that it is an intelligent design. However, the belief has not prevented them from keeping trying to find out the scientific reason(s) behind this fact.

E. The Only Scientific Solution

What we have here are both serious scientific problems and authority problems just like those in the geocentric era. The main scientific problem started several centuries ago from the long-time failure in finding the aether predicted to exist based on the teachings of PM to produce gravity, light, etc. Physicists began to develop no-aether postulations based theories violating the teachings and the laws of PM, which has been and still is the only scientifically proven science in physics. It is a **no-no** for scientists to do so, since science must be consistent and accumulative without contradiction. **The only scientific solution is therefore to go back to PM. It means that scientists have no choice but to continue their searching for aether**.

Chapter 6 – The Scientific Revolution

A. The Purposes of the Scientific Revolution

To separate Science from Religion and Other Authorities

The only scientific revolution in cosmology started in the16th Century by Copernicus and expanded to physics and all other fields of nature science by Galileo in the 17th Century. It is time to finish it now by identifying what real science is once for all. Astronomy began with a mathematical-model developed by Ptolemy based on the religious belief (unscientific postulation) that Earth was the center of the solar system. Even though the model (science) was wrong, mathematics could still make it fit to some observations to predict some planetary motions. Under the protection of both the religious belief of the public and the powerful church at the time, this religious pseudo-science era had lasted well over a thousand years. Copernicus found that the Sun, not the Earth, was the center of the solar system. However, due to the strong opposition from the church, his scientific views were shared by his followers in secret only. A century later Galileo proved Copernicus correct with the telescope he made himself and published his results. Although he was jailed for life, his publication had helped ending the religious pseudo-science era of Ptolemy.

The purpose of the scientific revolution therefore is to separate science from religions, whose beliefs are often unscientific. Science is the laws of the nature neither manmade nor alterable by man. Supporting real science may have to bravely fight against those authorities, which support wrong science and against real science. Science is very important to all nations and their people. All governments worldwide have their agencies to promote scientific research and to educate science. They must support real science to benefit their nations and people. They are however run by human and make mistakes. By supporting wrong science, they are not performing their duties and they should appreciate the effort and the help to correct their scientific mistakes.

To Discover Real Science

The development of experimental findings-based real science of physics also began from Galileo, who has led to the development of physics of mechanics (PM). Therefore, Copernicus and his followers of course including Galileo have started a scientific revolution not only to separate science from religion but also to search and develop real science based on experimental findings, observation, and logic reasoning instead of religious beliefs.

One of the important lesson scientists have learnt from the scientific revolution is that for a mathematical model or mathematical calculation to be scientifically correct, it must agree with all experimental findings. A mathematical theory or model like that of Ptolemy not based only correct scientific postulations cannot be correct. PM continues to demonstrate that mathematics is an important tool for science and its applications but is not science itself. PM has proven that (nature) science is to study objects having masses and the phenomena they produce. Neither an object nor its phenomenon can be a mathematical function.

Scientists have also come to agree with that science has only one correct answer only. It means that there can be only one correct or real science that must consistent without contradiction. Although wrong science may appear to agree with some experimental findings and/or observations, only real science can stand out in repeated tests supported by all experimental findings. One thing scientists have been doing in the past several centuries is to continue testing scientific laws and knowledge to determine among them which is the only correct or real science. They have found it (PM) but are losing it!

Therefore, another important purpose of the scientific revolution is to develop and to identify what the only real science is. After several centuries' concentrated effort in scientific R&D, there amazingly huge amount of experimental findings have been accumulated, which should be sufficient for identifying the only real science, which must be supported by all findings.

B. The Comeback of Religious Pseudo-Science

Although the scientific revolution in astronomy had been successful in changing geocentric to heliocentric view in solar system, it is far from completion. Cosmology remained to be the biggest mystery incomprehensible by mankind filled with wild imaginations and religious beliefs. For example, the steady-state planetary motions in the solar system had led to believe a steady-state universe without aether. Long-time failure in finding aether had further led to the development and acceptance of no-aether theories, which directly contradict with the teachings of PM. Over the past four centuries only PM has been repeated proven correct having all its postulations proven correct. Physicists therefore call it the determinant physics and core of physics. Meanwhile, several no-aether theories have been developed and accepted by the physics community. It should be inexcusable for any scientific community to make such a fatal scientific mistake as accepting two physics contradicting each other! Instead of scientific solving their serious scientific contradictions, physicists have been find such excuse as that they could not find aether and insisting that no-aether theories are compatible with PM. Having nothing to produce universal phenomena, physicists have tried very hard but could not scientifically understand their own no-aether theories, thus, forcing them to go back to religious beliefs. Before long, history had begun to repeat itself. Both the unscientific postulations and the mathematical models had made a strong comeback beginning in 18th century to interpret universal phenomena such as gravitational, electric, magnetic forces and light. Long-time failures in finding aether had been the main reason for more and more scientists to rebel against PM. Several coincidences also helped misleading physicists and cosmologists into religious beliefs, such as that mathematics was the Lord of (nature) science, again turning physics into a religion. For example: light speed happened to be the same as that of electromagnetic wave; Einstein's spacetime and curved spacetime could mimic the presence of an atmosphere aether (neutrinos) well. The phenomenal success of Newton's theory of universal gravitation had led the way for the comeback of no-aether postulations and mathematical model-based religious pseudo-science closely similar to that of Ptolemy. Einstein's (un)scientific revolution has finished the job of leading the world into a new era of religious pseudo-science leading all authorities to support religious pseudo-science while denying real science.

Since Einstein's "scientific revolution" a century ago, the scientific revolution started by Copernicus and Galileo not only had come to a full stop but also gone back to religious pseudo-science again! Now, the only real science is in grave danger again!

C. Fighting for Real Science

For the welfare and the future of mankind the scientific revolution started by Copernicus and Galileo must be continued and completed. To do so the scientific community, the government agencies supporting scientific research and education, and the general public need to realize that all no-aether theories including modern physics are wrong and that PM including the newly discovered new aether science is the only real science. It means that they need to read this book and compare the new aether science to what they have learnt and then take proper steps to correct the biggest scientific mistakes made over the past several centuries.

Already a century into an era of religious pseudo-science, it is very hard even to get someone to take the accusation seriously. Science belongs to everybody and everybody has the responsibility to protect it. The author would like to offer his experience in fighting for the real science and encourage everyone to continue the scientific revolution. After over four centuries' of R&D, the only real science for everything has now been discovered. The scientific revolution has been succeeded and the only thing we need to do is to end the era of religious pseudo-science!

The efforts of the author to publish new aether science, to gain recognition, and to apply for research funding in the past nine years have become the struggles to rescue our science. It is painfully clear that in the past century an all out war has been engaged by modern physics supporters to defeat PM supporters and they have won. Ever since they have been ruled the world of physical science with iron fist and continue to crack down hard on PM supporters. They have total control of government funding, scientific journals, R&D and education institutions and even have the support of the public worldwide. They have established unscientific rules and regulations to prevent aether-based concepts and discoveries from getting published and funded. Of course, to accomplish all these, they have not relied on science but on spreading religious beliefs and imaginative science fictions, which have gained popular supports.

The Corrupted Physical Science Journals

The author's efforts to publish his discoveries of new aether science have encountered unimaginable difficulties. All his submissions have been rejected by the editors of the journals of physical science and cosmology without giving any valid scientific reason. There have been some reasons vaguely stated by the editors either in the rejection letters or in their response to the author's insistent inquiries. For example, in their responses Physics Review Letters cited the lack of mathematical theory and Canadian Journals of Physics cited difference in philosophy.

Scientific journals have legal right to reject publishing of any submission with or without valid scientific reason. The journals of physical science and cosmology worldwide have long been abusing their freedom to support religious pseudo-science and to deny PM-based real science and findings!

National Science Foundation

As the major source for basic scientific research and education funding, National Science Foundation (NSF) plays a leading role in the scientific community. Unlike scientific journals and magazines, NSF should be officially responsible for maintaining a healthy scientific community to advance correct science since NSF spends tax dollars to promote the advancement of science. Besides, NSF should obey laws, for example, not to fund religion including religion-like activities. To perform its duties lawfully NSF needs to differentiate science from religion. The officers of NSF therefore should be familiar with and support fundamental scientific principles and need to have the ability to differentiate correct science from wrong science. It would be really bad for NSF to supports wrong science and to deny real science like what

NSF has been doing probably in the past century! To save real science, NSF must be convinced to support real science and to reject religious pseudo-science. The author has been doing so over the past nine years. Apparently, he needs lots of support and help! It would be a breakthrough only to convince NSF to judge the new aether science based on PM science not modern physics. Increased public requests that NSF sponsors public scientific debates on new aether science vs. no-aether theories is badly needed for rescuing real science. As shown below NSF at present stubbornly supports religious pseudo-science and denies real science!

Author's First proposal

The author's first research proposal was submitted in early 1999 entitled "A New Universal Theory – further development and evaluation". It was a very early version of the new aether science that showed scientific evidences to prove that the atmosphere of neutrino particles is the aether to produce both gravity and light. It still supported that the universe was uniformly expanding and suggested that the continuously expanding of the atmosphere of neutrinos was the cause. Among the four reviewers only two of them have raised scientific questions besides ridiculing the proposal with such languages as "lack of understanding of basic physics, etc.", which are given below:

Reviewer 1:

The proposal seeks further support to study a new approach to gravitation. Working from the perspective that special relativity and general relativity are not correct, and that field theory approaches to electromagnetism and gravity are unacceptable, the author attempts to explain a wide range of observations. The reviewer recognizes that one must constantly critique existing theories and propose new explanations; however, he feels that the current work falls short of existing theories by lacking internal consistency and predictive power.

The core of the proposed New Universal Theory is that gravitational interactions arise from the space subatomic particles, mainly neutrinos. Rather than accept the current field interpretation of gravity, the author looks for a physical origin for gravity. A wind of space subatomic particles, of which neutrinos play a large part, is thought to blow through the universe at very high speeds. This theory suggests that particles lose some kinetic energy as they pass through a massive body, and hence, the gravity wind flow out of the earth is weaker than the wind descending onto the earth from the heavens; this differential leads to an overall pushing of other matters towards the massive body.

The New Universal Theory is applied to two other physical situations. The first is the case of light and its bending around massive objects. Rejecting a field interpretation of electromagnetic waves, the proposal suggests that a suitable candidate for the ether is a wind of neutrinos, and that light waves are essentially mechanical in nature. Further, from the arguments given above demonstrating that a stronger gravity wind of neutrinos blows towards the massive object than away, one concludes that light will deflect towards a massive object because of the motion of the gravity wind through which the light travels. The New Universal Theory is also applied to cosmology; here, the neutrinos are assumed to have a natural dispersion which explains the expansion of the universe.

While recognizing that new formulations of gravity theories may be useful in defining a quantum theory of gravity or increasing our understanding of the world, the reviewer does not feel that the current proposal has great promise for adding to the body of knowledge. Granting for a moment that the New Universal Theory provides an internally consistent world view, it is not clear that the perspective of the theory has the potential to describe the wide range of observations currently explained by general relativity. In addition, the proposed theory does not (yet) give any sort of prediction, which can be tested to see if it fits our observations any better than existing theories.

Regarding the proposed New Light Theory, the reviewer wonders if the author is familiar with the work of many prominent scientists at the end of the 19th century on the properties of ether. A solid check of internal consistency of the New Universal Theory would be to show that a collection of neutrinos satisfies the properties which the ether must posses to transmit electromagnetic waves.

The reviewer sees that the proposal has the possibility of being an important step in our understanding of the world. However, the current proposal is lacking in two ways. First, the reliance on neutrinos as a gravity wind by the New Universal Theory seems to limit the power of the theory to explain all of the observations explained by existing field theories. Second, the proposed theory has yet to make any predictions which could be tested for accuracy. The reviewer feels that both of these issues should be addressed by the author of the proposal.

Reviewer 2:

The following report is based on the March 26, 1999 revised version of 'New Universal Theory', which is part of NSF proposal PHY-9907068. In support of his proposal, Dr. Tsau presents a 'New Universal Theory', which is based on the assumption that 'essentially undetectable' subatomic particles occupy all of space. These 'space subatomic particles' (SSPs) are used by Tsau to develop new theories of gravity, light, and the expansion of the universe. Tsau claims that the existence of SSPs disproves Einstein's theory of special relativity and extends the general theory of relativity, by providing an underling mechanism that is responsible for the transmission of light and gravitational interactions. I am not convinced, however, by Tsau's arguments, and I do not recommend that the proposal be funded.

1. Although Tsau claims that a 'gravity wind' of SSPs is the mechanical explanation of the gravitational fields associated with massive objects, this "New Gravity Theory' cannot explain, for example, how the earth could remain in orbit around the sun for billions of years without slowing down due to collisions with the SSPs. (More SSPs will hit the front of the earth than the back as the earth moves in its orbit, creating a force that opposes the motion.) In addition, Tsau states that 'for very dense matter, the strength of gravity is affected by its size only and no longer affected by the amount of its mass.' This is direct contradiction to experimental observations which show that the gravitational held of a massive object is proportional to its mass and not its size.

2. When discussing the Doppler effect of light, Tsau states that 'if there is no transmission medium, conventional Doppler effect no

longer exists. A Doppler-like effect can only be observed by changing the speed of emitted particles'. This statement is incorrect. In special relativity, even though one assumes the absence of a transmission and an absolute, observer-independent speed of light, one is still able to predict the existence of a Doppler effect when there is relative motion between the receiver and source of light. More over, this predicted Doppler Effect has been verified, for example, by the red-shifted light we observe from receding galaxies. Tsau also fails to realize that the Doppler effect for sound is fundamentally different from that for light ---i.e., the Doppler effect for light depends ONLY on the relative motion of source and receiver, while that for sound depends on whether the source or receiver is moving with respect to the transmission medium.

3. According to Tsau's 'New Light Theory', light travels 'at a constant speed … (only) relative to its transmission medium, SSP. If light is emitted by an (outer) galaxy which is approaching or receding from us, the speed of light would be correspondingly larger or smaller. In making such statements, Tsau is incorrectly applying Galilean addition of velocities to relativistic objects (such as light). The fact that the Galilean addition of velocities formula fails for relativistic objects (and should be replaced by the Lorentzian addition formula) is verified experimentally everyday in high-energy accelerators. Tsau seems to have neglected this observational fact when formulating his theory.

4. Tsau claims that current theories of gravity do not obey the law of conservation of energy. This statement is incorrect. Newtonian gravity and Einstein's theory of general relativity DO obey the law of conservation of energy (albeit only globally in general relativity). Granted, Newton's theory of universal gravitation and Einstein's general relativity do not provide the type of mechanistic description of gravitation that Tsau wants, but I do not agree with his statement that general relativity is only a 'pure mathematical model without physical meaning'. A theory (like general relativity) which agrees with experiment and explains HOW objects move in response to gravitational interactions (but does not necessarily explain WHY) is better than a theory which purports to answer WHY but does not agree with experiment.

For these and other reasons not mentioned here, I do not recommend that the proposal be funded.

Comments of the Author

Reviewer 1 contradicted himself regarding many aspects such as the "internal consistency" of the theory. He pointed out some of the important values of the proposal but very reluctantly. Reviewer 2 is a typical defender of the religious pseudo-science to keep real science from publishing and getting funded. He simple uses the scientifically unproven knowledge of no-aether theories as the scientific criteria to deny PM based real science. The "other reasons not mentioned here" are that his prestigious job and his NSF-funded projects were based on modern-physics knowledge. He wanted "business as usual" at the cost of science. NSF does not give proposer a chance to answer reviewers' comments. However, at the time the author did not have all the answers to reviewers' questions even when he was given the chance to rebut them. Now all the answers to both reviewers can be found in this book.

Author's 2nd proposal to NSF

NSF does not accept re-submissions of rejected proposals or similar ones. It was only because that the author wrote to the Director of NSF Dr. Colwell pleading her to save our science. Dr. Dehmer, Director of the Division of Physics, responded suggesting that I submit a proposal. My second proposal "New Aether Science" was submitted in early 2003 and was rejected in June with the following major reasons.

Reviewer 1: It is so differently at odds with the current standard view of physics, I think it would be inappropriate for the NSF to support this work at any level until the PI has demonstrated its credibility by generating a series of published accounts of the work in the main stream refereed scientific literature.

Reviewer 2: The proposer has no track record of published research in gravitation physics in refereed publications.

Reviewer 3: There is no published record of the investigator's previous work in the field. The ideas presented in the proposal are purely qualitative and no quantitative underlying theory apparently exists at present. ...at present string theory offers an entire gamut of new alternatives...

Comments of the author

Again, it is a blind rejection of PM-based real science by all reviewers. There is no evidence that any of the reviewers has gone through the proposal and judge its scientific merits. They have rejected it simply because it was PM based. This time they all emphasized on publishing record as Dr. Dehmer mentioned in his letter given below.

Despite what the scientific community support, NSF has the official responsibility to support real science. It was the officers of NSF, who religiously supported modern physics, had chosen reviewers having the same views and goals to protect modern physics to review and reject PM-based scientific physics. The following letters from NSF reveal their biased views and stands against real science.

Letters from NSF Officers

1. From Dr. Dehmer, Director of Physics, dated on 2/14/03

Dear Dr. Tsau: Thank you for your 9 January 2003 letter to Dr. Rita Colwell expressing your concerns about the progress of modern physics and transmitting your book, "Mysteries of Nature." I am impressed with your dedication to science and your independent work. Your revival of the aether concept in a modern context is, indeed, quite a departure from current ideas. Many individuals with unorthodox ideas have had difficulty gaining acceptance in the mainstream of thought. However, the ideas are tested by criticism and experimental tests; and the best concepts, models, and theories generally emerge over time.

You obviously feel that you are in such a position now, having non-standard ideas and experiencing difficulty gaining acceptance. You appeal to the NSF because we sponsor basic research and have a responsibility for the health of science. However, our role is not to pass judgment on new ideas. Our role is to sponsor the most meritorious research efforts. The merit of proposals is determined by peer review, which is the best tool we have to assess the quality of proposals.

We will be happy to consider any research proposal that you may wish to submit. However the burden of demonstrating the correctness and promise of your ideas lies with you. This is normally done through peer-reviewed publication.

Author's Comments

Dr. Dehmer agrees that NSF has a responsibility for the health of science by sponsoring the most meritorious (scientific) research efforts. He however called the newly discovered aether and new aether science, which are supported by both all experimental findings and PM, the unorthodox ideas, which is unacceptable by the modern-physics knowledge, the "current ideas" or "the mainstream of thought". Like all other physicists he must agree that modern physics is compatible with PM since PM is the fundamental physics taught in schools and colleges. However, PM is unacceptable to him and NSF by applying its knowledge to the universal phenomena and cosmology. What Dr. Dehmer and NSF are actually doing is to deny proven correct (real) science to support modern physics or the mainstream which represents theories of relativity-based modern physics and modern cosmology.

By submiting the new aether science to the evaluation system based on the knowledge of modern physics and its formal science philosophy, Dr. Delmer did not give it (newm aether science) a chance to be (scientifically) judged as he promised (the ideas are tested by criticism and experimental tests; and the best concepts, models, and theories generally emerge over time). The proposal was seeking funding and the support of NSF to have the new aether science evaluated by the scientific community. Yet NSF has chosen a few reviewers who religiously support modern-physics views to deny the chance for making the new aether science known to the public, evaluated and debated against modern physics and modern cosmology.

Presently physics has two scientifically incompatible cores, PM, the proven correct science and no-aether theory. PM has to be the only real science. Yet, physicists including Dr. Dehmer and their scientific authorities have been trying very hard to support no-aether theory and to deny PM. Upon discovery of aether and new aether science, they must return to real science now!

2. Letter from Dr. Berger, Program Director, Gravitational Physics, dated on 3/14/2003

Dear Dr. Tsau: While some aspects at the frontiers of physics are controversial, the core of physics is based on agreement between theory and experiment. Even "revolutions" in physics such as relativity and quantum mechanics did not destroy the previous core but rather demonstrated that it had a limited region of validity. The "new" physics was adopted into the core when it became clear that it, too, was supported by experiments in the regime where the "old" core did not apply. A new revolution must also be consistent with previous experiment and should itself be testable by existing or new experiments. You should note that, in addition to textbook experiments, the current core of physics is supported by the successful functioning of modern technology. An interesting example of this is the Global Positioning System that has effects of special and general relativity included in its hardware and software. Thus, every time a position is correctly determined, special and general relativity are validated.

Comments from the author:

The scientific explanation of the need to adjust the frequency of the laser used by Global Positioning System is given in this book. Dr. Berger's letter explained in more detail about the official stand of NSF on the scientific issues of physical science. NSF supports modern physics (theory of relativity and quantum theory) as a newly added core of physics besides PM, the original core of physics. It means that NSF has accepted all the fundamental postulations of modern physics to be true despite that they have not been proven correct experimentally. More importantly, due to the incompatibility between the two cores of physics, choosing one would against the other. NSF chooses its reviewers to use the knowledge and the formal science philosophy of modern physics as the scientific criteria to judge scientific proposals. What it means is that although PM is still the core physics, NSF actually is against PM, both its scientific teachings, particularly its aether view and its scientific method that relies on experimental findings only not mathematical-models to discover scientific origins, which are physical objects not mathematical functions. Science must be consistent and we can not support two completely contradicting sciences, since one of them must be wrong. What NSF has been doing is to replace PM with modern physics not really to support both.

Similar to scientific journals NSF uses peer-review method to run its scientific research funding programs. It appears to be the best tool but no method is fool proof. It certainly will not work when the people administrating it religiously support wrong science and are not impartial. It is apparent that the reviewers NSF had chosen for the author's proposals are all staunch supporters of modern physics sharing the same views and beliefs with the officers of NSF. NSF needs unbiased officers to run the office!

The officers of NSF must be capable of differentiating between science and pseudo-science and between science and religious beliefs. They are the ones, not the reviewers they choose, responsible for supporting correct science only.

NSF is the most powerful leader of the scientific community responsible for the health of the scientific community. Yet, NSF has long been supporting religious pseudo-science against real science. Public insistence of having a responsible and law-abiding NSF is the key to rescue real science.

Legal Challenge

Financially supporting scientific R&D and science education have been the major efforts of all nations, since it is a scientifically and technologically competitive world. To stay as the leader of the world United States spends billions of dollars per year in scientific R&D and education.

Meanwhile, U. S. constitution separates government from church and forbids its government from supporting religious activities.

The U. S. Supreme Court is particularly well known in protecting science and in striking down religious pseudo-science. The famous ruling of The Supreme Court for evolution against creation suggests that the court has recognized that there could be only one correct science. Although both creation and evolution are largely unproven theories, supporting evolution means supporting the overall scientific evidences based on PM. Creation is not a matters-balance theory and is not compatible with PM. Historically, U. S. courts and the justice systems worldwide have long been relying on physics of mechanics (PM) based science, such as finger print, DNA, etc. daily for ruling cases.

What the U. S. law needs but has not done is to have a legal definition for nature science! Without one, the scientific community has the freedom to define their own science. For example, they can define a religion to be science. Without one, U. S. government has been support more than one nature sciences contradicting one another but only one of them can be the real science. Worse yet, U. S. government has been supporting religion in the name of supporting science, thus, has violated its own constitution! In recognition of the need of striking a balance between the desire for vigorous public debate and the fear of governmental intrusion into that debate, the First Amendment purposely leaves open the possibility that all viewpoints may not be heard! Applying this law to science would mean that the only real science may not be supported by the government! The spirit of this law can not be applied to nature science to destroy real science and it is the responsibility of The Supreme Court to prevent this from happening!

With the huge amount of scientific evidences accumulated over the past four centuries, it is time for The U. S. Supreme Court to recognize that PM is the only real nature science and no-aether theory including modern physics and modern cosmology is not. At least, PM should not be denied by the government agencies such as NSF.

The following is a letter the author wrote to the Supreme Court on 26 January, 2006 hopefully to open a legally battle to rescue real science. Although no response has been received, it is worthwhile to share it with readers. The author hope that the real science supporters will continue legal challenges until it is saved.

The Honorable Judges of the Supreme Court
Supreme Court of the United States
One First Street, NE
Washington, DC 20543

RE: The Urgent Need of a Legal Interpretation on Science to differentiate Real Science from Religious Pseudo Science by the Supreme Court

U. S. Constitution separates Government from church. Government sponsors basic scientific research and supports science education to benefit the nation and mankind. It is however difficult to separate science from religions (church). Since the beginning, for example during the Geocentric Era, "science" was inseparable from religion and was a religious pseudo-science (RPS) based on religious beliefs, which were unscientific postulations.

The urgent need of a legal definition on science is from the very fact that the world has again been in a RPS era for a century already! It means that we have been educated to believe that RPS, also known as modern physics, is the real science and our children are learning the concepts of RPS since grade schools through colleges. RPS wastes our heavy investments in scientific research and education, for example with our tax dollars, kills young minds, and harms our welfare and future! It should be illegal and unconstitutional for our government to promote and to finance RPS while denying real science to hurt the general public!

There have been many legal cases regarding separation of government from church and a large portion of them involving separation of science from religion. The genuine efforts from the courts across the nation to protect science and to separate it from church are very clear. Without a clear legal definition on real science, however, courts have no effective legal tool to protect real science and they have already lost a big battle to corruptions for a century already. The following facts are presented to support my urgent request from the Supreme Court.

As described in the attached book entitled "Discovery of Aether and its Science" the only repeatedly proven correct physics (science) in the past four centuries is the physics of mechanics (PM) that Galileo has led to develop. Since science can have only one correct answer, it must be consistent and accumulative without contradiction. Scientists must obey the teachings of proven correct science to ensure consistency in science. According to the teachings of PM all phenomena, which have forces and/or energies, are produced by objects. For example, the presence of an atmosphere of air (objects) on Earth produces such worldwide phenomena as sound, atmospheric pressure, physical and chemical changes, temperature changes, etc. PM predicts that there should be an atmosphere of some kind of object(s) (aether) taking up all the space of the universe to produce such universal phenomena as light, gravity, etc. However, long-time failure in finding aether has corrupted scientific community to develop and to accept no-aether postulations-based theories (NAPT), such as Newton's Theory of Universal Gravitation and Einstein's Theories of Relativity. Using postulations violating PM, NAPT should be incompatible with PM, thus, scientifically wrong. Based on both religion-like beliefs and using mathematical models to replace PM-based real science, NAPT have turned into a new version of Geocentrism; its over-a-thousand-year-long RPS era had been ended by the scientific revolution led by Copernicus and Galileo.

In summary, over the past several centuries, the scientific community has made the biggest scientific mistake by accepting both PM and NAPT, which contradict with each other, to be the cores of physics. In fact, the scientific community has officially abandoned PM to support NAPT since Einstein's (counter) scientific revolution a century ago. Although physicists continue to admit that PM is the core physics, they religiously believe that all universal phenomena are mathematical equations (the so-called formal-science philosophy) instead of objects from the teachings of PM. They even believe that the universe is a mathematical equation, such as the already well known String Theory under development supported by NSF and our government directly and/or indirectly. Using the above (religious-like) philosophy and other unscientific knowledge of NAPT as scientific standards, the scientific authorities, such as NSF, have effectively denied PM-based real science for a century already!

The discovery of both aether and its new science (Chapter 1 of the attached book entitled Discovery of Aether and its Science) is the major scientific breakthrough since Galileo by expanding PM to the science of everything. It has uncovered a simple consistent coherent scientific understanding of the universal phenomena and the universe and has further conclusively proved PM while disproved all NAPT. It now clearly shows that all the universal phenomena produced by aether (the atmosphere of neutrinos) have been mystified by NAPT into RPS. For example, light is misbelieved to be a mathematical function known as electromagnetic wave (not an object) and a object-free pure energy (according to PM only objects produce energies) called photon believed to have such unscientific properties as traveling at an absolute constant speed without transmission medium, thus, leading to beliefs that time and dimensions are not absolute. Gravitational force is mystified to be both an acting-from-distance attraction force among all heavenly bodies (PM teaches that objects only produce pushing force acting only by contact) and even object-free curved spacetime. Also, the phenomenon of universal red-shift of star lights discovered by Hubble, which is also produced by aether, has been mystified to create Big Bang Theory that the universe was born from a big bang of a "singularity" about 15 billion years ago and has been expanding ever since leading to the scientifically wrong modern cosmology.

Like ancient Catholic Church, one of today's powerful authorities supporting RPS is National Science Foundation (NSF)! For nearly a century, NSF has been using the (unscientific) knowledge and the formal-science philosophy of modern physics, the modern RPS, as scientific standards to support (fund) the research and the education of RPS while denying real science. In the past six years, the new aether science has been discovered and presented twice to NSF but has been repeatedly rejected using the same RPS standards. The facts about NSF's blind supporting RPS and denying real science are given in the Chapter 3 of the attached book.

There can be only one real science, which is accumulative, consistent and coherent without contradiction and is supported by all experimental findings. Only real science can stand out in the presence of tremendously large amount of experimental findings accumulated over the past four centuries, which is the PM. It is very clear now that PM is the real science and that all NAPT are not. For example, even if distorted spacetime appeared correct in interpreting gravitational force, it can not explain other forces. Yet, the teaching of PM that objects produce act-on-contact pushing force scientifically and mathematically explains all forces including gravitational force. In another example, modern physics postulates that in nuclear reactions a small amount of mass is lost turning into object-free "pure energy" obeying Einstein Equation $E= mc^2$. This interpretation contradicts with the well established PM-based knowledge that in all chemical and physical reactions mass is balanced and that object-free energy is indefinable scientifically and, thus, does not exist. The PM-based interpretation that agrees with all experimental findings is that in nuclear reactions some neutrinos, which have masses, are released. The explosive expansion of the released neutrinos produce huge amount of (nuclear) energy. Also, only the PM-based interpretation that light is the particle wave of neutrinos is consistent with the well established science of sound.

The discovery of aether and its new science has again conclusively confirmed the correctness of PM and disproved all NAPT. Aether (the atmosphere of neutrinos) was also found to be the common scientific origin of all universal phenomena since its unique physical properties can explain and predict the existence of all of them. Besides, the atmosphere of neutrinos is also the only thing that fills all space of the universe constantly replenished by all stars. Therefore, a simple consistent coherent scientific understanding of the universe including all of its phenomena has presented itself upon the discovery of aether and it is consistent with the teachings of PM. It is very clear now what we have been learning (from all NAPT) are religious beliefs and mathematical fantasies and we have been in a RPS era for a century already.

Wrong sciences are easily spotted and corrected upon getting new scientific evidences against them. A RPS however is based on religious beliefs and its believers are essentially blind to scientific disproof. For centuries the public including scientists have been educated to believe the correctness of such NAPT as Newton's Theory of Universal Gravitation, that light is a mathematical equation called electromagnetic wave, Force Field Theory,

Einstein's Theories of Relativity, Quantum Theory, Big Bang Theory, etc., they are incapable of believing otherwise since they do not really scientifically understand them. Like religions they are simply believers and followers. Governments, particularly their scientific agencies and laws are responsible for preventing the public from the harm of corruptions. There is however too much at steak for those people in charge of such government agencies as NSF to admit their mistake and to end their corruptions!

By supporting RPS and denying real science U. S. governments including its agencies have been harming science, thus, harming her people for nearly a century already! Without a clear legal definition of "science" or what is "real science", U. S. law can not legally differentiate real science from religion (RPS) to protect real science and U. S. Constitution of separation of government from church. Further more, the First Amendment purposely leaves open the possibility that all viewpoints may not heard to strike a balance between the desire for vigorous public debate (on social issues) and the fear of governmental intrusion into that debate. Unfortunately, the First Amendment has been abused to treat NAPT-based RPS as a legal option of science and by legally denying real science as a viewpoint that may not heard! Unlike social issues having many options, the only real science must be legally heard and protected.

Unlike social issues, there can be only one real science that is accumulative, consistent, and supported by all experimental findings. Our government and its agencies must protect and support real science not RPS to protect the public and its constitution. Therefore, a definition of real science based on the above cited features of real science is needed from the Supreme Court to legally differentiate real science from religion and RPS. Scientific research in the past four centuries has produced and accumulated huge amount of experimental findings, only PM stands out supported by all of them. It is time to legally define that PM is real science. In fact, the PM-based science such as finger prints and DNA have long been admitted in U. S. Courts and in courts' rulings based on that there is only one correct answer in science. In the famous case of evolution vs. creation, the Supreme Court's ruling in favor of evolution indicates that PM has been chosen as real science against religious teachings or RPS. Even though evolution is still largely an unproved theory, its postulations are consistent with the teachings of PM such as biology and chemistry while the religious teachings of creation are not.

PM and NAPT can not both be real science. NAPT are based on postulations violating the teachings of PM and therefore are incompatible with PM. The failures in finding aether for several centuries have gradually misled scientists to accept NAPT and to naively believe that NAPT are still compatible with PM but are valid only to universal phenomena. Meanwhile, without finding aether, PM has no scientific answer to universal phenomena. Therefore, the major scientific contradiction between PM and NAPT has been hidden while PM-based science has been denied by both religious beliefs in NAPT and by the lack of PM-based scientific answers to universal phenomena. Upon finding aether and new aether science, there is no more scientific reason for the government and its scientific agencies to continue to support NAPT and their RPS.

It is urgent for Supreme Court to speak out defining what real science should be and to rule on which PM or NAPT has been proven to be the real science and which the government should and should not support to save our science and to end the century old RPS era.

Made in the USA
Monee, IL
08 November 2026

67382840R00079